99-WWF-4

SENSING AND CONTROL SYSTEMS: A REVIEW OF MUNICIPAL AND INDUSTRIAL EXPERIENCES

by:

Robert D. Hill, Ph.D., P.E.

Robert C. Manross, P.E.

EMA, Inc.

Elizabeth V. Davidson, P.E.

Elizabeth Davidson & Associates, Inc.

Tony M. Palmer

Maureen C. Ross, P.E.

Instrumentation Resting Association

Stephen G. Nutt, P. Eng.

XCG Consultants, Ltd.

2002

Water Environment Research Foundation

The Water Environment Research Foundation, a not-for profit organization, funds and manages water quality research for its subscribers through a diverse public-private partnership between municipal utilities, corporations, academia, industry, and the federal government. WERF subscribers include municipal and regional water and wastewater utilities, industrial corporations, environmental engineering firms, and others that share a commitment to cost-effective water quality solutions. WERF is dedicated to advancing science and technology addressing water quality issues as they impact water resources, the atmosphere, the lands, and quality of life.

For more information, contact:
Water Environment Research Foundation
601 Wythe Street
Alexandria, VA 22314-1994
Tel: (703) 684-2470
Fax: (703) 299-0742
www.werf.org
werf@werf.org

This report was co-published by the following organizations. For non-Subscriber sales information, contact:

In the U.S.:
Water Environment Federation
601 Wythe Street
Alexandria, VA 22314-1994
Tel: (800) 666-0206
Tel: (703) 684-2452
Fax: (703) 684-2492
www.wef.org
pubs@wef.org

International:
IWA Publishing Alliance House, 12 Caxton Street
London SW1H 0QS
UK
Tel: +44 (0)20 7654 5500
Fax: +44 (0)20 7654 5555
www.iwapublishing.com
publications@iwap.co.uk

Library of Congress Catalog Card Number: 2002108845
Printed in the United States of America
WERF ISBN: 1-893664-54-6 IWAP ISBN: 1-84339-656-4 WEF ISBN: 1-57278-205-6

EMA Inc.

ACKNOWLEDGMENTS

Report Preparation

Principal Investigator:
Robert D. Hill, Ph.D., P.E.
EMA, Inc.

Project Team:
Robert C. Manross, P.E.
EMA, Inc.

Elizabeth V. Davidson, P.E.
Elizabeth Davidson & Associates, Inc.

Tony M. Palmer
Maureen C. Ross, P.E.
Instrumentation Testing Association

Stephen G. Nutt, P.Eng.
XCG Consultants, Ltd.

Project Subcommittee

Robert A. Reich, P.E. (Research Council Liaison)
DuPont Company

Richard Haugh (Subcommittee Chair)
Carollo Engineers

Richard Finger
Seattle County

Michael K. Stenstrom, Ph.D., P.E.
University of California – Los Angeles

Z. Cello Vitasovic, Ph.D., P.E.
Camp Dresser & McKee

Expert Advisory Panel

John F. Andrews, Ph.D., P.E.
Professor Emeritus, Rice University

Francis Greg Shinskey
Process Control Consultant

Allan M. Springer, Ph.D.
Miami University

Michael W. Sweeney, Ph.D., P.E.
Louisville Metropolitan Sewer District

Robert E. Werner, P.E.
District of Columbia Water & Sewer Authority

Environment Research Foundation Staff

Deputy Director, Research: Charles I. Noss, Sc.D.
Research Program Director: Jeff C. Moeller, P.E.
Project Manager: Mary Strawn

ABSTRACT AND BENEFITS

Abstract

The objective of this research was to assess and document state-of-the-art of wastewater treatment plant sensing and control systems to discover successful practices that can be replicated by other facilities interested in automating. The research was based on the literature, the experience of the project and advisory teams, and onsite surveys. The project focused on the best examples of applied sensors, control strategies, and computerized process control in wastewater treatment plants.

An instrumentation and control survey designed to assess the current state of automation showed that most instruments are used for the simpler measurements, and most respondents justified installing automation systems because of cost savings, though less than 10% of the facilities had data demonstrating these savings. In addition, the survey showed that wastewater treatment facilities that do use automation and remote monitoring during at least part of the day do achieve great cost savings.

The 25 field surveys done to gather site-specific details provided useful data on both successful and unsuccessful automation practices. The project team found a large number of control strategies not commonly documented in current literature. Critical components were documented, and when costs and savings were known, these values were also reported.
The report lists the 10 factors important for control system success, which resulted from factors identified by attendees at a WEFTEC 2000 workshop, combined with data from the field surveys. Almost all these factors relate to organizational and management rather than technical issues.

This report also identifies needs that must be met if the full potential of wastewater treatment sensing and control systems is to be realized. While some of these needs are technical, perhaps the most important need is for continued demonstration of the real-world benefits of automated control. Though this project identified many instances where automation has realized significant benefits, many more cases comparing the costs and performance of a WWTP before and after implementation of a comprehensive system will be needed to convince the wastewater industry of the future for sensing and control systems.

Benefits

- Provides the results of an extensive instrumentation and control survey of 110 wastewater treatment facilities.
- Provides a compilation of control strategies for major process areas from 25 field surveys.
- Provides a large body of information on both successful and problematic practices that will give managers confidence in making the decision to automate.
- Provides a discussion of the 10 Keys to Success found to most influence the success of wastewater plant automation projects in actual practice.
- Provides a discussion of trends in the instrumentation and control practice.
- Provides a discussion of the future needs of automation, instrumentation, and control for wastewater treatment plants.

Keywords: Automation, instrumentation, control, hardware, software, strategy

TABLE OF CONTENTS

Water Environment
Research Foundation

LIST OF TABLES

LIST OF FIGURES

LIST OF ACRONYMS

CMMS	Computerized Maintenance Management System
CPI	Chemical Process Industry
DAF	Dissolved Air Flotation
DCS	Distributed Control System
DO	Dissolved Oxygen Concentration
DOC	Dissolved Organic Carbon
DRP	Enterprise Resource Planning
FDDI	Fiber Distributed Data Interface
GC	Gas Chromatograph
HART	Highway Addressable Remote Transducer
HMI	Human-Machine Interface
LAWPR	International Association on Water Pollution Research, now the International Water Association (IWA)
I&C	Instrumentation and Control
I/O	Input/Output
ISA	Instrument Society of America
IT	Information Technology
ITA	Instrumentation Testing Association
IWA	International Water Association
LEL	Lower Explosion Limit
mgd	million gallons per day
MIS	Management Information System
ORP	Manual of Practice
PID	Oxidation-Reduction Potential
PLC	Proportional-Integral-Derivative
PV	Programmable Logic Controller
SPC	Process Variable
SRT	Statistical Process Control
TOC	Solids Retention Time or Sludge Age
TOD	Total Oxygen Demand
Y2K	Year 2000
U.S. EPA	United States Environmental Protection Agency
UV	Ultraviolet Radiation
WEF	Water Environment Federation
WERF	Water Environment Research Foundation
WWTP	Wastewater Treatment Plant

EXECUTIVE SUMMARY

ES-1 Purpose of Study and Project Approach

The objective of this study was to assess and document the state-of-the-art of wastewater treatment plant sensing and control systems through an examination of the literature, the experience of the project and advisory teams, and onsite surveys. By one definition, "state-of-the-art" means the current condition of the industry today. By another definition, "state-of-the-art" means the highest level of achievement or best practices. This project addressed both issues but focused on the best examples of applied sensors, control strategies, and computerized process control in wastewater treatment plants. This report includes a large body of information on successful and problematic practices that will enable users in other facilities to replicate the documented successes for comparable treatment processes using state-of-the-art monitoring and control.

ES-2 Instrumentation and Control Survey

An instrumentation and control survey was used to assess the current state of automation. A total of 110 treatment facilities representing 45 utilities and industries responded to the survey. Results of the survey are discussed in Chapter 2, but some of the findings were:

- Most of the instruments in use today are for the simpler measurements such as flow, pressure, level, position, and temperature. Many facilities also use analytical instruments for measurement of dissolved oxygen and chlorine residual, although fewer use them for automatic control. Few facilities use the more sophisticated analyzers.

- The majority of respondents installed automation systems for their cost savings in labor, energy, and materials. However, less than 10% of the facilities had quantified data demonstrating these savings.

- Many wastewater treatment facilities are attended 24 hours per day. However, a number of facilities use automation and remote monitoring during at least part of the day. These results may have been somewhat biased by the 41 surveys from the City of Houston, which monitors most of its facilities remotely.

ES-3 Field Survey

The instrumentation and control survey was structured to allow easy comparison among facilities but had limited capacity for gathering subjective and site-specific details about facilities. The field survey provided a mechanism to gather these additional details from specific facilities through one-on-one interviews with field staff. The project team conducted 25 site surveys in the United States and Canada. The standardized control strategy summaries from the field survey are discussed in Chapter 3.

Perhaps the most useful data captured during the field surveys are the many case histories of successful and unsuccessful automation practices. The project team found a large number of control strategies not commonly documented in current literature. Chapter 4 documents these control strategies on a unit process basis. Critical components, as reported by the owner, are documented. When costs and savings were known, these values are also reported. However, the majority of utilities documented neither costs nor savings. Many of the case histories are illustrated with a process and instrumentation diagram.

The case histories are perhaps the most important product of this study in that these practices can often be directly adapted by other utilities.

ES-4 Industrial Knowledge

One initial objective of the project was to research automation in other industries and bring this knowledge base back to the wastewater industry. This portion of the project was perhaps less successful than other parts, since many industrial contacts could not talk about the technologies they were using because of competitive pressures. Also many of their successes are somewhat unique and have a very narrow range of applicability.

As part of the industrial input for the project, the project team interviewed Mr. F. Greg Shinskey, a well-known industrial process control expert. Excerpts from that interview are documented in Appendix C. Two of the most relevant points Mr. Shinskey made were:

- When retrofitting a plant, particularly from manual to automatic, all of the old controls need to be taken away in order for the operators to accept the new way of running a process. Not doing this makes it possible for operators to fall back on the old way, putting the new system at risk of not being accepted and used.

- In petroleum and chemical projects the process designers and the instrumentation and control (I&C) designers work together in a team from the beginning of the project. Mr. Shinskey observed that in the wastewater industry process design often does not include I&C as an integral part of the job.

ES-5 Keys To Success

Attendees at a WEFTEC 2000 workshop identified factors they considered contributory to control system success. A tally of these success factors, combined with information from the facilities visited during the field surveys, resulted in a top 10 list of success factors or Keys to Success. The list starts with the most frequently occurring success factor (having a system advocate) to the 10th most frequently occurring factor (setting goals). The list was validated during the WEFTEC 2001 workshop where 39 attendees (93%) agreed with the list, while 3 (7%) disagreed.

The list contains mostly organizational and management related issues, rather than technical ones. The list is not complete, nor does it guarantee system success when followed, but it does indicate some of the common success factors. The Keys to Success are discussed in detail in Chapter 5.

ES-6 Needs

This project identified numerous needs that must be met if the full potential of wastewater treatment sensing and control systems is to be realized. While some of these needs are technical in nature, perhaps the most important is for the continued demonstration of the real-world benefits of automated control. Despite the perception that automation of WWTPs has not achieved its full potential, this project identified many instances where automation had realized significant benefit. However, many more cases comparing the costs and performance of a WWTP before and after implementation of a comprehensive automation system will be needed to convince the industry of the potential of WWTP sensing and control systems.

Chapter 1.0

INTRODUCTION

1.1 Motivation

The justification for using instrumentation and control systems has remained consistent for over two decades. The 1978 WPCF Manual of Practice (MOP) 21 states "....increased emphasis on treatment efficiency and increased labor costs required better and more sophisticated plant sensing and control." This MOP project was motivated by a need to improve the performance record of process control instruments. Subsequent publications to get "the word out" on how to implement successful automation systems have been produced by the the US EPA, WERF, and WEF through updates to its MOPs. The latest such manual is the 1997 special publication entitled *Automated Process Control Strategies.* Practical how-to-automate articles are also a favorite of technical book publishers and journal editors. All these efforts have helped improve the acceptance and performance of sensors, control strategies, and computer process control systems in WWTPs.

Much of the work on technology transfer has been the collaborative effort of designers, plant operators, and suppliers. In parallel with this knowledge sharing, traditional hardware and software products have improved in reliability, ease of use, and price/performance, as well as an infusion of new technologies not available even a few years ago. Advances in networking/digital interfaces, wireless communications, and information data warehousing have positioned the wastewater industry to evolve applied instruments and control strategies in WWTPs to a new level in contributing to improved treatment efficiencies and labor cost management.

The foundation of any automation strategy, however, requires reliable, repeatable on-line sensor technology. Associated with each measured variable is cost. In addition to the obvious purchase and installation costs is a commitment to ongoing service to keep the sensor in good working order. This means an expectation or value judgment is associated with the data generated by an instrument and the needed maintenance effort. In the past, there have often been unrealistic expectations associated with the care and maintenance of instrumentation.

Competitive pressures create the need to consider increased use of automatic process control based on advanced strategies and on-line monitors. The large body of information on both successful and problematic practices compiled in this study will help make managers confident in their decisions to automate.

1.2 Research Supports Continued Application of Automation Technology

Despite demonstrated improvements and optimism for new applied technology, automated systems are not problem-free. Close examination of problem sensors and control loops has uncovered reasons for less than satisfactory performance. One case involved erratic performance of a chlorine residual analyzer. Analysis found inconsistent calibration procedures and location in a poorly mixed area following the point of chlorine addition. Other problems are maintenance related such as a faulty temperature compensation circuit on a sonic level sensor in a flow measurement application. The result of the correction and testing was a significant reduction in the reported effluent flow from the plant.

A persistent problem when applying control strategies in WWTPs is oversized process equipment operating at the low end of its control range. When an analysis of the total loop is performed, often the control equipment is working on the edge, a difficult to impossible application to effectively deliver regulatory control of the process. In other cases, multiple levels of control (local, area, and central) or automatic backup build in layers of complexity that contribute to the instability of a process control system.

These are only a few of the real world examples where sensors and automation are reported to be problematic. Unfortunately these situations foster concerns about building a dependency on instruments and process control systems for the continuous operation of complex treatment plants.

It is, therefore, timely for a project of this type to compile and update the wide range of experience using actual case studies to determine what works and to assess how the wastewater industry is progressing in the use of this technology to achieve more cost-efficient operations.

1.3 Research Objective

The objective of this research was to assess and document the state-of-the-art of wastewater treatment plant sensing and control systems through examination of the literature, the experience of the project and advisory teams, and onsite surveys. The phrase "state-of-the-art", however, can have two different meanings. By one definition, "state-of-the-art" means the current condition of the industry today. By another definition, "state-of-the-art" means the highest level of achievement or best practices. This project addressed both issues but focused on the best examples of applied sensors, control strategies, and computerized process control in WWTPs. This report includes a large body of information on both successful and problematic practices that will enables users in other facilities to replicate the documented successes for comparable treatment processes using state-of-the-art monitoring and control.

The investigations showed:

- what works,
- why it works,
- what is needed to keep it working,
- what tangible benefits it provides, and
- how to repeat the documented success.

Conversely where applied automation has not worked, the study showed:

- what does not work, and
- why it does not work.

1.4 The Business Case for Automation

Whether automation practices are successful or not, supporting data and experience are required, at least in part to dispel myths about instrumentation. By setting factual expectations WWTPs can continue moving toward highly efficient treatment operations aided by applied automation technology.

Fifteen to 20 years ago, when many WWTPs were undergoing new construction or expansion, instruments and control systems were often incorporated just because something was needed to operate the plant. In many cases the cost justification for the automation work was based on estimates with little supporting data. This lack wasn't critical at the time, since the automation systems normally represented a small percentage of the total construction budget.

Today, however, most of these older systems have exceeded their technological life and need replacement, so cost justification plays an important role in deciding to proceed. The results of this study identify the benefits derived and provide support to others building the business case to justify the replacement or upgrade of their current systems.

1.5 Final Research Report Compiles Up-To-Date Experiences

Today more than ever the industry is focusing on the cost to deliver wastewater services. Continuous process and discrete manufacturing industries have used analytical sensors and advanced control strategies in their product production to improve performance, reduce waste, provide consistent quality, ensure worker safety, and increase profit margins. This project also examined technology advances in industry that showed promise to enhance wastewater treatment efficiency.

The final compilation of the project research represents a comprehensive report on current state-of-the-art sensors, control strategies, and process control that deliver improved WWTP performance when correctly applied. How to do it, what to avoid and benefits to be derived are part of the information provided. Emerging technology trends were examined with suggestions for pilot projects to evaluate these technologies to continue the improvement in operations cost effectiveness.

1.6 Research Approach

Figure 1-1 illustrates how the real experiences of numerous sources were consolidated into this final report.

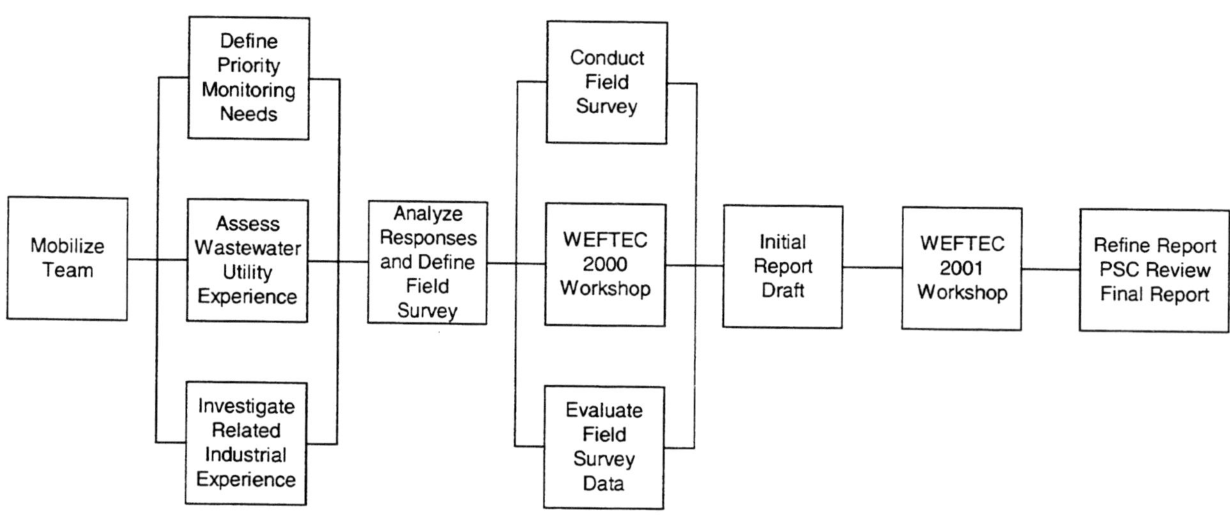

Figure 1-1. Project Approach Overview

1.6.1 Assess Utility Experience

The first three parallel tracks of the project consisted of defining a priority monitoring list, assessing the wastewater utility experience through a survey, and investigating related industrial experience. The priority monitoring list was developed through a literature search to determine the most common variables for monitoring treatment plant performance. Its main use was to help define the survey instrument.

The survey to assess wastewater utility experience was based on the priority monitoring list and previous survey experience. It collected data to document the practical side of the monitored applications, maintenance requirements, and expected costs. The seven-page survey is attached in Appendix A. Results of the survey are presented in Chapter 2. The survey specifically addressed:

- sensors and instruments in use, performance, maintenance requirements, cost;
- control strategies;
- computerized control system architecture (hardware, software, and communications);
- overall system effectiveness; and
- potential system improvements.

For-profit industries offer related monitoring successes that can be transferred to water utilities. These industries long ago demonstrated the value of using real-time monitoring and control to achieve improved product quality, worker/plant safety, and efficiency of operation. The trade publications and trade associations were searched for relevant experiences, practices and guidelines to add to the practical worth of the report.

1.6.2 Field Surveys

A survey instrument was designed so that all field investigators collected complete sets of data including sensors in use, sensor performance, sensor costs, sensor maintenance requirements, control strategies, control system equipment, and other factors as deemed appropriate. Unfortunately not all of the facilities surveyed had all of the desired information. A copy of the field survey form appears in Appendix B.

The field surveys were critical to the success of the project in that they reflect real-world experiences from day-to-day use. Each survey identified the successful (or unsuccessful) application of sensor technology, control strategies, and computer control systems to improve plant performance and improve plant efficiency through reductions in power, chemicals, labor, and other costs.

Each survey included as a minimum (where available):

- formal interviews with operations and maintenance personnel and plant managers;
- review of written documentation of control strategies including process and instrumentation diagrams (P&IDs);
- instrumentation installation, maintenance, operations, training, and safety practices;
- review of plant performance, compliance, and cost data;
- review of maintenance records and reports;
- control system staffing requirements; and
- walk-through tour of the facility to document processes, instrument condition, installation details, and other pertinent site conditions.

Twenty-five facilities in the United States and Canada were selected as sites for an in-depth field survey. Sites were selected primarily for their success in applying control technology as discovered in the background tracks. Consideration was also given to geographic distribution and industrial facilities.

Equipment manufacturers, model numbers, and other data were sometimes collected during the site surveys to aid in the analysis of the survey data. However, specific suppliers and model numbers are not reported in this project report. To avoid implied or explicit endorsement or criticism of specific products or proprietary technologies, this report focuses on the applied technology rather than the manufacturer or product.

The results of the field survey are presented in Chapters 3 and 4. Results were consolidated into summary guidelines for process performance monitoring and control. For each measurement application, installation, typical maintenance requirements, and services were identified along with an estimate of life-cycle costs when available. Both successful and unsuccessful control strategies were cataloged. Process control systems were critiqued to evaluate their success in performing required functions.

1.6.3 WEFTEC Workshops

Two WEFTEC workshops were developed as part of this project. Participants in the first, an invitation-only workshop organized for WEFTEC 2000, included consultants, utility managers, industry, suppliers, and others known to the project team and the project subcommittee. Twenty people participated. Preliminary results of the project were presented along with the overall project plan. The participants were solicited for process control strategies and success factors. The workshop participants largely developed the list of success factors discussed in Chapter 5.

The WEFTEC 2001 workshop presented the results of the I&C survey and seven process control strategies. The participants themselves were solicited for 17 additional control strategies, several of which are discussed in Chapter 4.

1.7 Why Isn't Automation Used More Universally in the Wastewater Industry?

The International Association on Water Pollution Research (IAWPR) held the first large conference specifically on instrumentation and control of wastewater systems more than 26 years ago in London, England (1973). IAWPR, now the International Water Association (IWA), continues to hold similar conferences around the world every four years with the latest in Malmö, Sweden in June 2001. The Water Environment Federation (WEF) has held three specialty conferences (1993, 1995, and 1997) on instrumentation and control. Every WEFTEC annual conference for the past 15 years has included several technical sessions and at least one workshop on instrumentation and control.

Virtually every paper and presentation given at these conferences and workshops over the past quarter century has discussed the significant advantages of wastewater facilities automation through instrumentation and control. Some of the advantages claimed are improved performance, additional treatment capacity, reduced labor requirements, and electrical and chemical savings. While some of these papers are based on theoretical considerations, many present irrevocable evidence of these advantages by reporting on actual performance and financial data from years of use in full-scale facilities.

The surveys conducted for this project documented the fact that many wastewater treatment facilities utilize little automation. In those facilities utilizing automation, only about half of the control loops run in the automatic mode. With such a history of evidence pointing to the many advantages of automation through instrumentation and control, why is it that many facilities still have very little automation and depend mostly on manual labor? Why do only half the control loops work?

While some technical barriers remain to fully implementing automation, most of the barriers are the result of characteristics of the wastewater treatment processes themselves and the characteristics of the organizations that design, own, and operate the plants, according to Hill (1999) and Vitasovic (2001).

1.7.1 Characteristics of the Process

A variety of characteristics make automation difficult. Some of these are:

- large and somewhat unpredictable changes in inputs;
- complex kinetics of processes;
- difficulty of measuring important process variables;
- interactions between process units;
- conservative design traditionally done by steady-state approximation; and
- low-cost product.

While each of these characteristics is important, other industries with similar characteristics have been able to move beyond them. Perhaps the most hampering of the characteristics is the tradition of conservatively designing wastewater treatment plants with steady-state approximations. Many plants are actually intentionally designed with little control so that the operators can't make process adjustments that might upset the plant.

1.7.2 Characteristics of the Organization

Some of the characteristics of organizations that design, own, and operate wastewater treatment facilities make automation difficult. These are:

- large and somewhat unpredictable changes in inputs;
- funding cycles favoring the conservative approach;
- lack of profit motive for many POWT facilities;
- constrained rules for POWT facilities, especially for procurement;
- difficult metrics for success; and
- a high penalty for failure and low reward for success.

The wastewater industry, as a whole, has been very conservative and risk adverse for many years. Automation was often considered a risk with a high penalty for failure and little reward for success. Consequently, many facilities were built with little automation. The competition from private operating firms, however, has changed this dynamic. Now many facilities are turning to automation to improve efficiency and effectiveness and increase their competitiveness.

1.7.3 Hope for the Future

Despite the many obstacles, this study reports on many innovative and practical uses of instrumentation and automation in wastewater treatment. Many have increased the effectiveness of treatment. Others have slashed costs by reducing labor, energy, and chemical requirements. All have contributed to improved treatment effectiveness and efficiency and serve as examples of what can be accomplished in the future.

Chapter 2.0

INSTRUMENTATION AND CONTROL SYSTEMS SURVEY

2.1 Introduction and Background

The first task of this project, to assess the success of process monitoring and automation technologies in wastewater treatment, included an in-depth survey of the instrumentation and automation systems used to control wastewater treatment plants.

The survey specifically asked questions regarding online instrument analyzers, sensors, and process control systems using a seven-page survey form (see Appendix A). The survey was designed to investigate the ability of instrumentation and control systems to provide cost-effective process control through accuracy and reproducibility.

The first task survey was directly distributed to all Water Environment Research Foundation (WERF) and Instrumentation Testing Association (ITA) members (approximately 80 for ITA and 220 for WERF). In addition, ITA provided the survey on its Web site with hyperlinks from the WERF Web site.

A total of 110 treatment facilities representing 45 utilities and industries responded to the first task survey. Of the 110 surveys, 41 were filled out on City of Houston wastewater treatment plants. This large number of Houston plants appears to have biased the survey results only in the area of unattended operation (see Figures 2-1 and 2-2). Using the results of the first survey, WERF project team members identified approximately 15 treatment facilities to be further investigated in the second task of the project, field surveys.

Results of the first task survey follow.

2.2 General Treatment Plant Information and Demographics

Approximately 96% of the survey responses were received from municipal wastewater treatment plants, 15% of which have advanced wastewater (tertiary) treatment. The remaining responses (4%) were received from industrial facilities. Only 5% of the surveyed treatment facilities are privately owned, primarily the industrial treatment facilities; corresponding percentages (95% and 5%) have public operational responsibility versus private operational responsibility. Plant hours of attended operation are shown in Figure 2-1 as compared to average daily flow. Most plants are attended 24 hours per day, while the second largest group is attended 8 hours per day.

This may reveal that the plants in the second group have a sufficient level of automation to require only 8 hours per day of attended operation. As mentioned earlier, these data may have been skewed by the inclusion of 41 plants from the City of Houston, which monitors many of its facilities remotely.

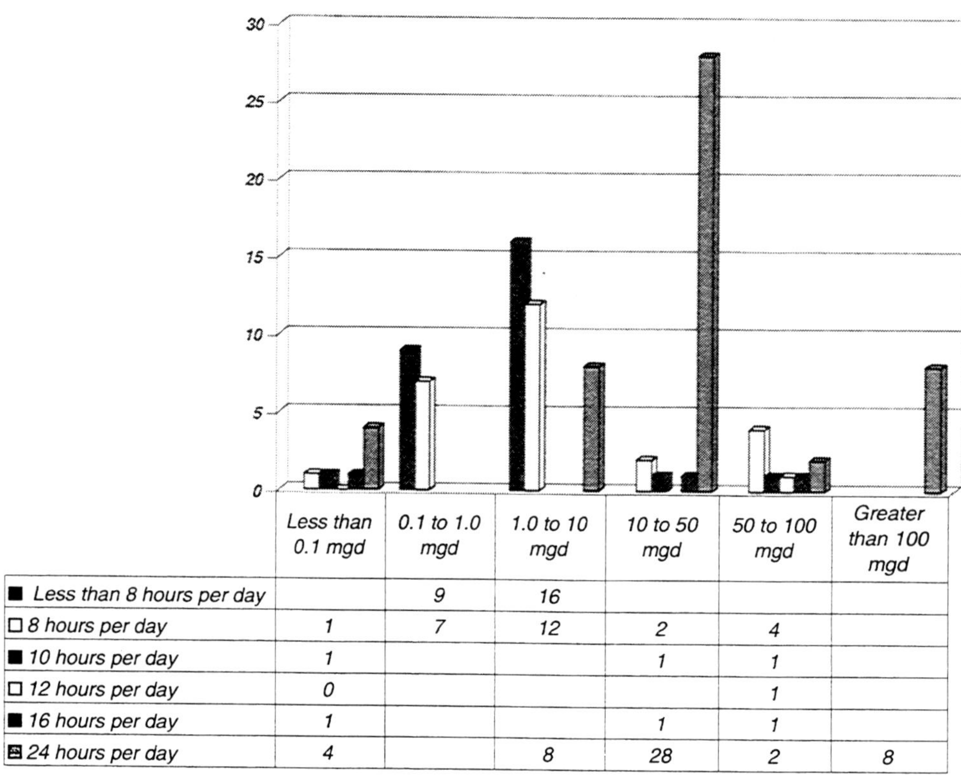

	Less than 0.1 mgd	0.1 to 1.0 mgd	1.0 to 10 mgd	10 to 50 mgd	50 to 100 mgd	Greater than 100 mgd
■ Less than 8 hours per day		9	16			
□ 8 hours per day	1	7	12	2	4	
■ 10 hours per day	1			1	1	
□ 12 hours per day	0				1	
■ 16 hours per day	1			1	1	
▤ 24 hours per day	4		8	28	2	8

Figure 2-1. Plant Hours of Attended Operation as Compared to Average Daily Flow

The distribution of plant days of attended operation for varying flow capacities is shown in Figure 2-2. These data may have been skewed by the inclusion of 41 plants from the City of Houston, which monitors many of its facilities through automation of critical processes and by a central SCADA system.

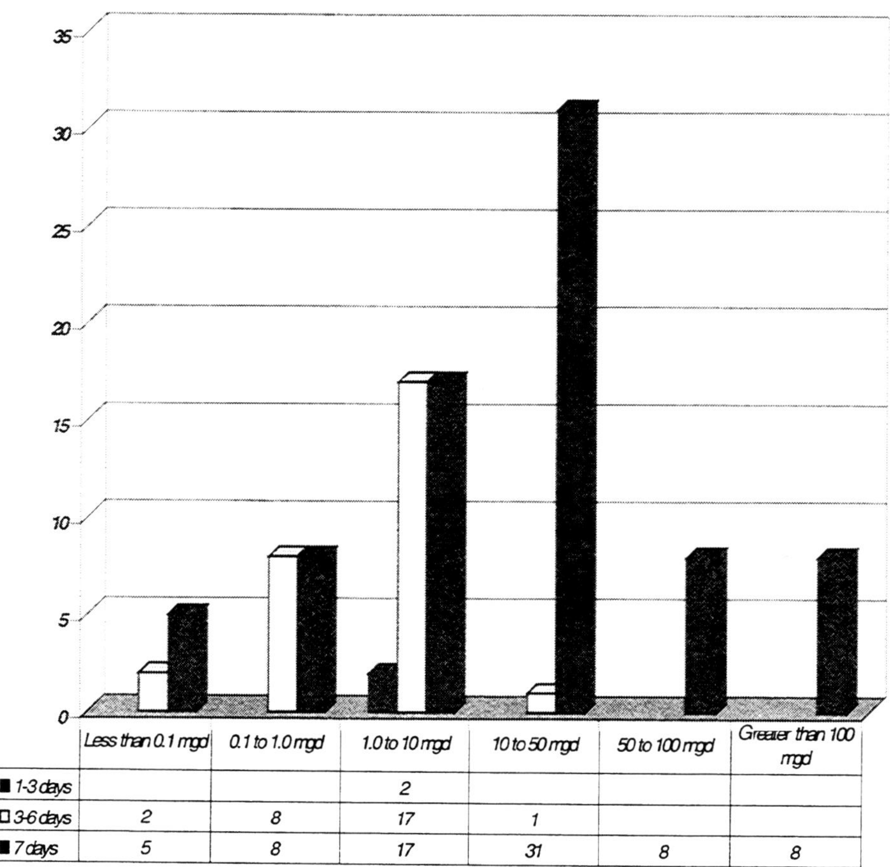

	Less than 0.1 mgd	0.1 to 1.0 mgd	1.0 to 10 mgd	10 to 50 mgd	50 to 100 mgd	Greater than 100 mgd
■ 1-3 days			2			
□ 3-6 days	2	8	17	1		
■ 7 days	5	8	17	31	8	8

Figure 2-2. Plant Days of Attended Operation as Compared to Average Daily Flow

Savings were reported as the greatest driving force for automating treatment facilities (for both staff time savings and energy and materials savings), as shown in Figure 2-3. Improved process control and early warning of process upsets were the next highest reported driving forces for automating. Correcting for varying conditions, was reported as the lowest motivation for automating, which ironically is what automation does best.

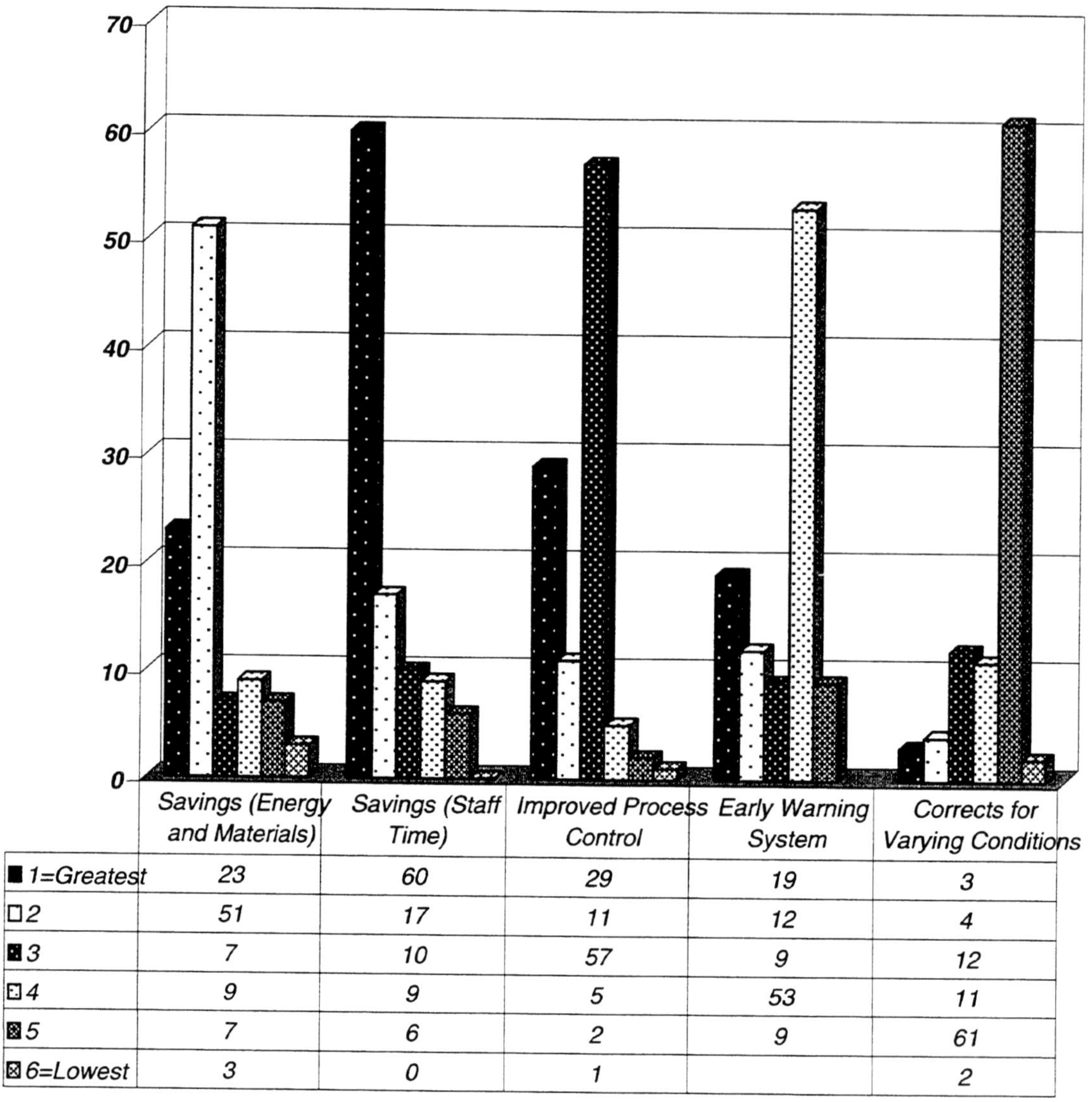

	Savings (Energy and Materials)	Savings (Staff Time)	Improved Process Control	Early Warning System	Corrects for Varying Conditions
1=Greatest	23	60	29	19	3
2	51	17	11	12	4
3	7	10	57	9	12
4	9	9	5	53	11
5	7	6	2	9	61
6=Lowest	3	0	1		2

Figure 2-3. Driving Forces for Plant Automation

Interestingly, the survey showed that many smaller treatment facilities have at least some automation and control. One plant of less than 1 mgd had over 1,000 I/O points indicating a significant amount of automation.

As Figure 2-4 shows, there are multiple distributions for I/O control points. For the smaller treatment facilities, there is a distribution for 0-100 I/O control points. For the larger treatment facilities, another distribution is shown for 1,000 to 5,000 I/O control points.

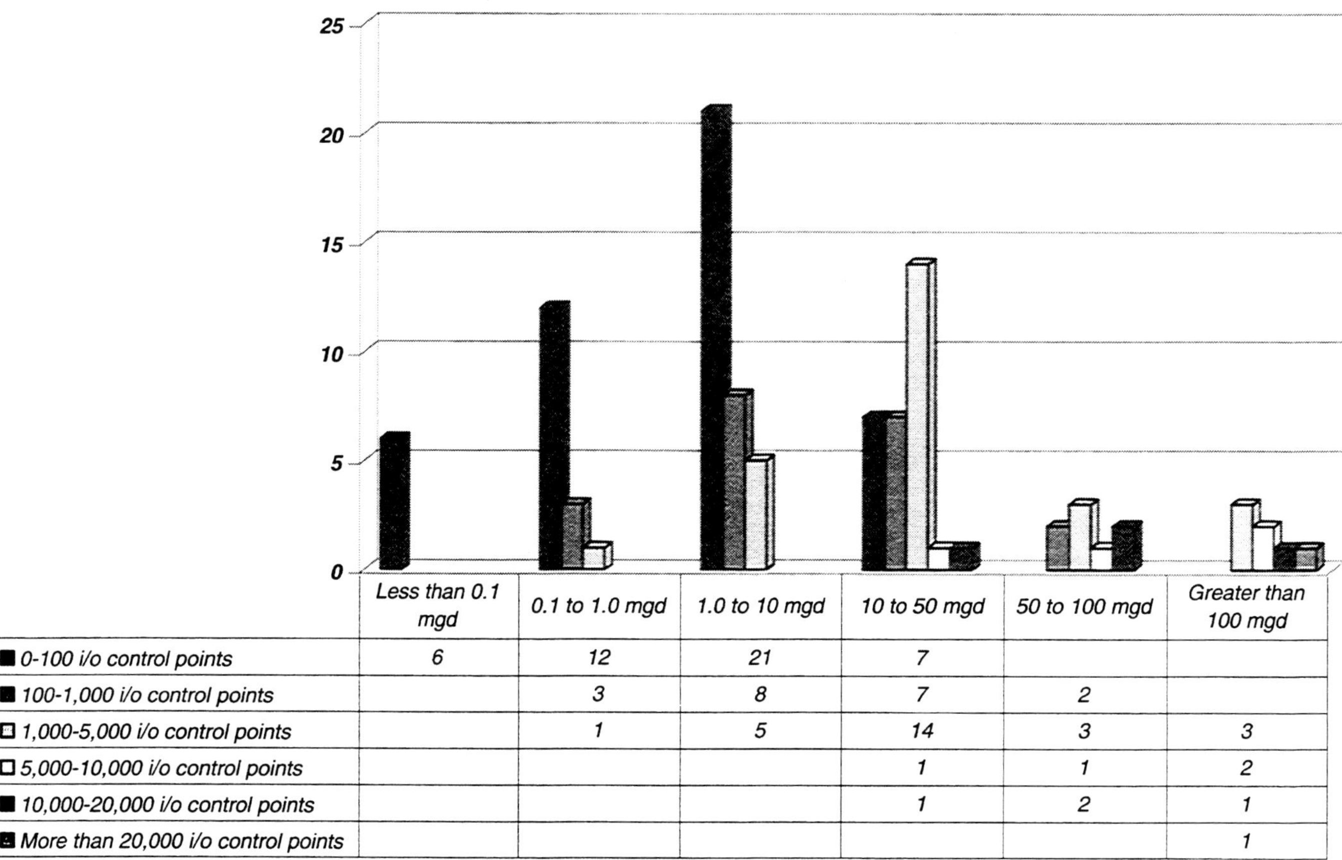

	Less than 0.1 mgd	0.1 to 1.0 mgd	1.0 to 10 mgd	10 to 50 mgd	50 to 100 mgd	Greater than 100 mgd
0-100 i/o control points	6	12	21	7		
100-1,000 i/o control points		3	8	7	2	
1,000-5,000 i/o control points		1	5	14	3	3
5,000-10,000 i/o control points				1	1	2
10,000-20,000 i/o control points				1	2	1
More than 20,000 i/o control points						1

Figure 2-4. Number of Control System I/O Points by Size of Plant

Almost all of the surveyed treatment facilities reported that control strategy documentation is available, as shown in Figure 2-5.

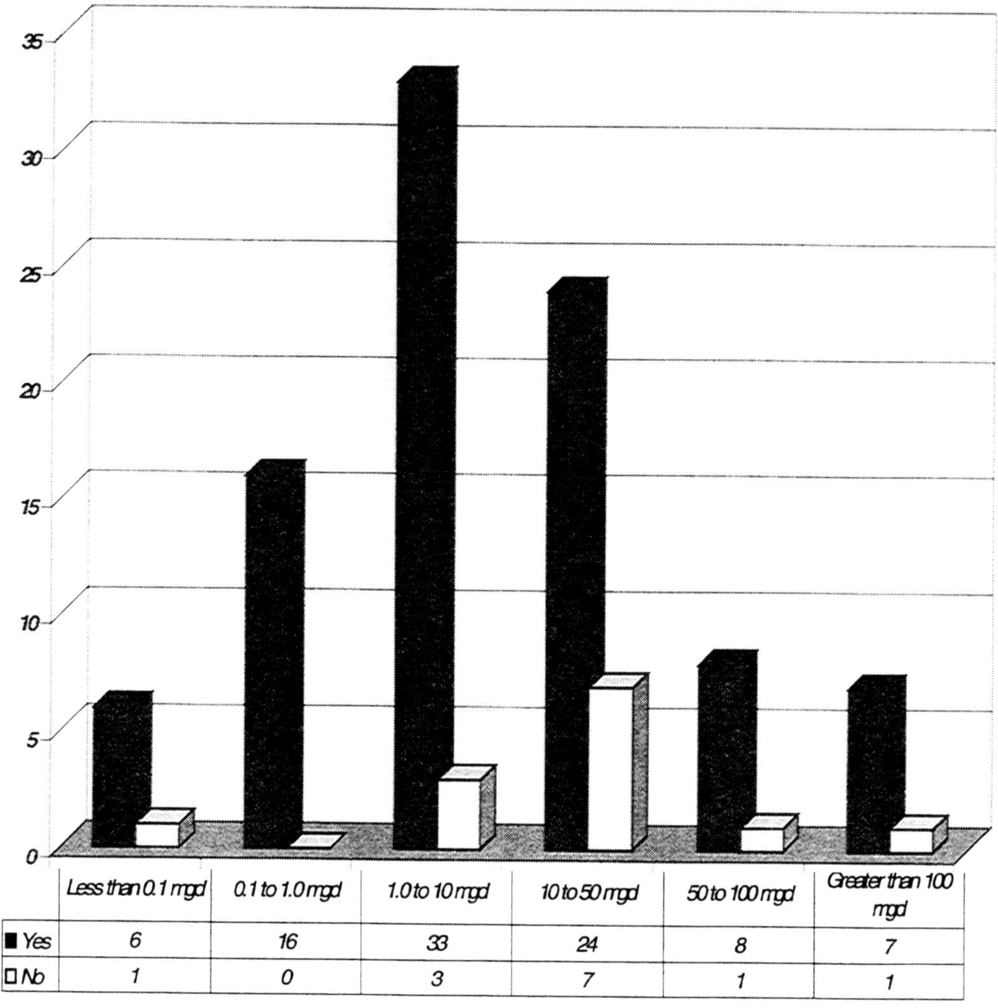

	Less than 0.1 mgd	0.1 to 1.0 mgd	1.0 to 10 mgd	10 to 50 mgd	50 to 100 mgd	Greater than 100 mgd
■ Yes	6	16	33	24	8	7
□ No	1	0	3	7	1	1

Figure 2-5. Facilities with Control Strategy Documentation by Size of Plant

It is interesting to note that the majority of the treatment facilities listed cost savings as a reason for automating, yet no cost data is tracked, as shown in Figure 2-6. Without accurate tracking of cost savings, justification for further investments in automation cannot be substantiated.

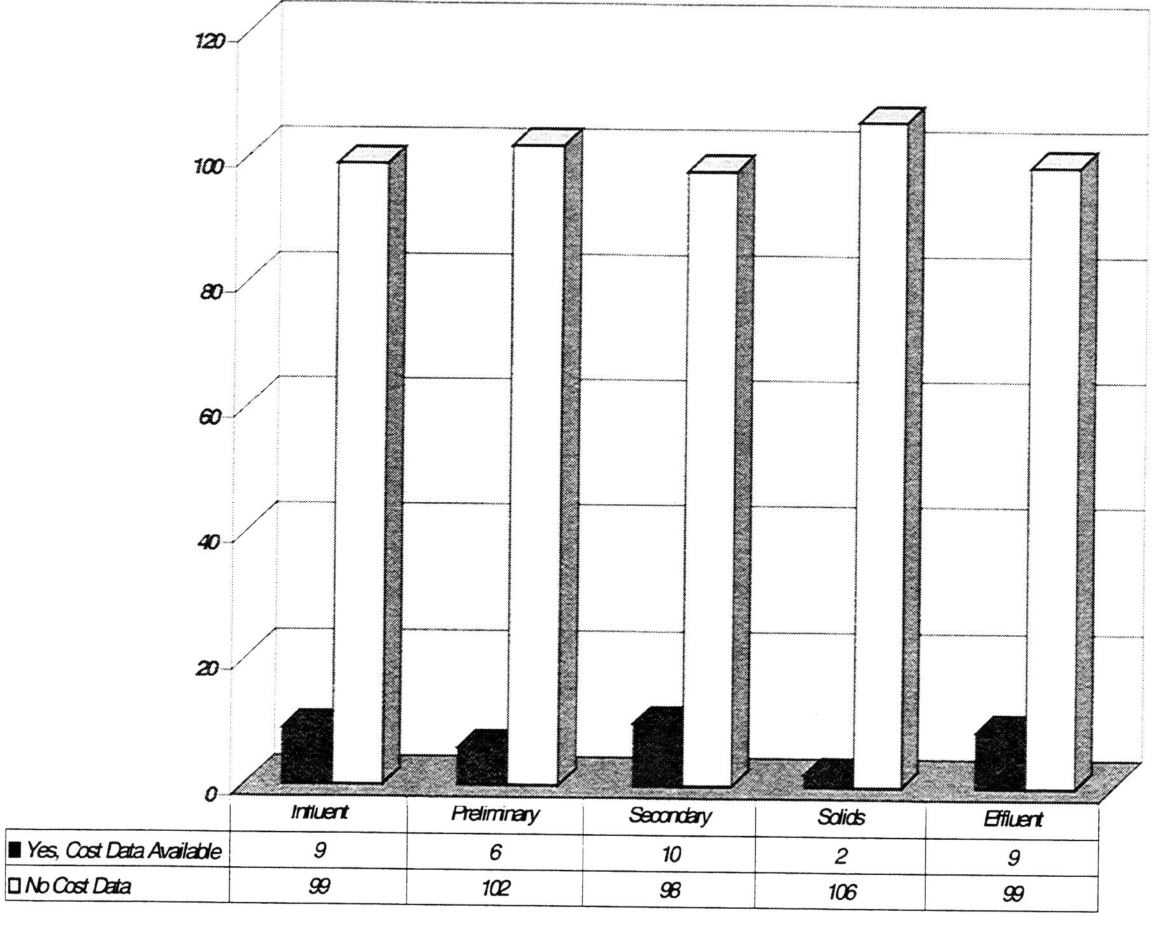

	Influent	Preliminary	Secondary	Solids	Effluent
■ Yes, Cost Data Available	9	6	10	2	9
□ No Cost Data	99	102	98	106	99

Figure 2-6. Availability of Cost Savings Data by Process Area

2.3 Sensor Use

Figure 2-7 shows the instrumentation used for influent, lift stations, and collection systems. A significant number of treatment facilities reported using the following instrumentation:

- level,
- position,
- liquid Flowmeter,
- lower explosion limit (LEL) and hydrogen sulfide (H_2S),
- accelerometers, and
- Watt meters.

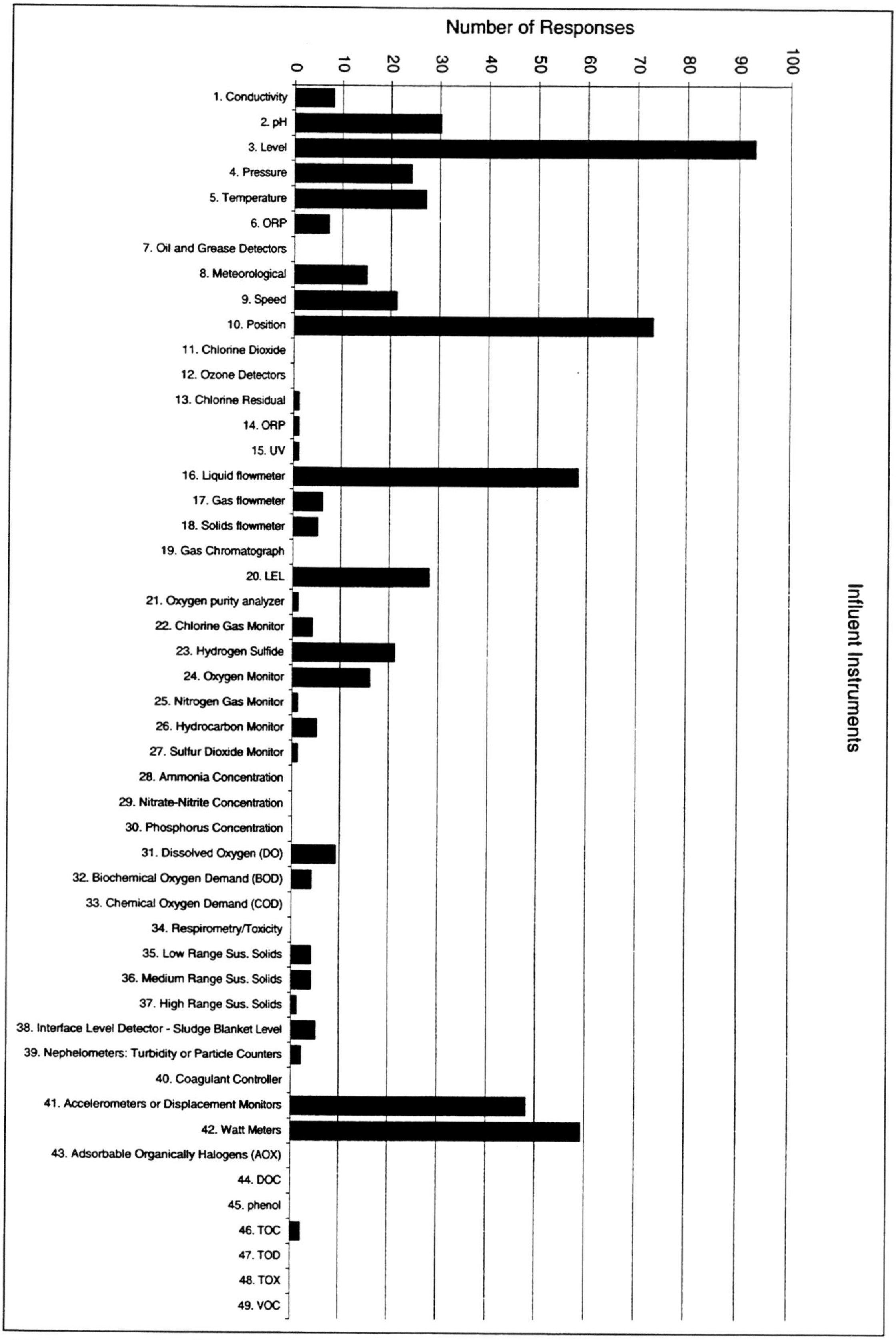

Number of Responses

0 10 20 30 40 50 60 70 80 90 100

Influent Instruments

1. Conductivity
2. pH
3. Level
4. Pressure
5. Temperature
6. ORP
7. Oil and Grease Detectors
8. Meteorological
9. Speed
10. Position
11. Chlorine Dioxide
12. Ozone Detectors
13. Chlorine Residual
14. ORP
15. UV
16. Liquid flowmeter
17. Gas flowmeter
18. Solids flowmeter
19. Gas Chromatograph
20. LEL
21. Oxygen purity analyzer
22. Chlorine Gas Monitor
23. Hydrogen Sulfide
24. Oxygen Monitor
25. Nitrogen Gas Monitor
26. Hydrocarbon Monitor
27. Sulfur Dioxide Monitor
28. Ammonia Concentration
29. Nitrate-Nitrite Concentration
30. Phosphorus Concentration
31. Dissolved Oxygen (DO)
32. Biochemical Oxygen Demand (BOD)
33. Chemical Oxygen Demand (COD)
34. Respirometry/Toxicity
35. Low Range Sus. Solids
36. Medium Range Sus. Solids
37. High Range Sus. Solids
38. Interface Level Detector - Sludge Blanket Level
39. Nephelometers: Turbidity or Particle Counters
40. Coagulant Controller
41. Accelerometers or Displacement Monitors
42. Watt Meters
43. Adsorbable Organically Halogens (AOX)
44. DOC
45. phenol
46. TOC
47. TOD
48. TOX
49. VOC

Figure 2-7. Instrumentation Used for Influent, Lift Stations, and Collection Systems

Instrumentation used for the preliminary and primary treatment process areas is shown in Figure 2-8. Here many of the surveyed treatment facilities reported using:

- pH,
- level,
- position,
- liquid flowmeter,
- LEL and H_2S, and
- Watt meters.

For secondary treatment, a significant number of facilities shown in Figure 2-9 reported using the following:

- pH,
- level,
- pressure,
- temperature,
- speed,
- position,
- liquid flowmeter,
- DO,
- accelerometers, and
- Watt meters.

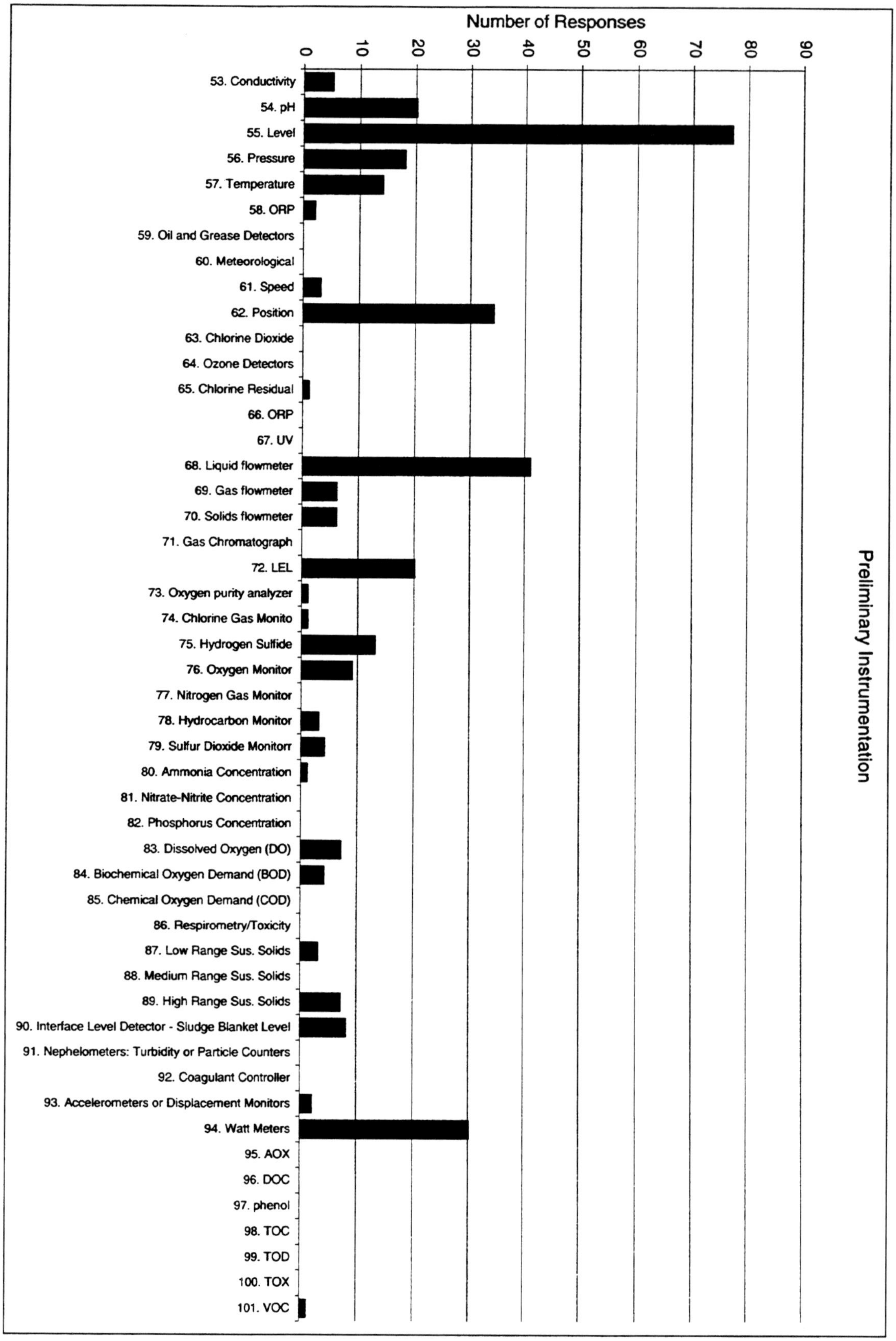

Figure 2-8. Instrumentation Used for Preliminary and Primary Treatment

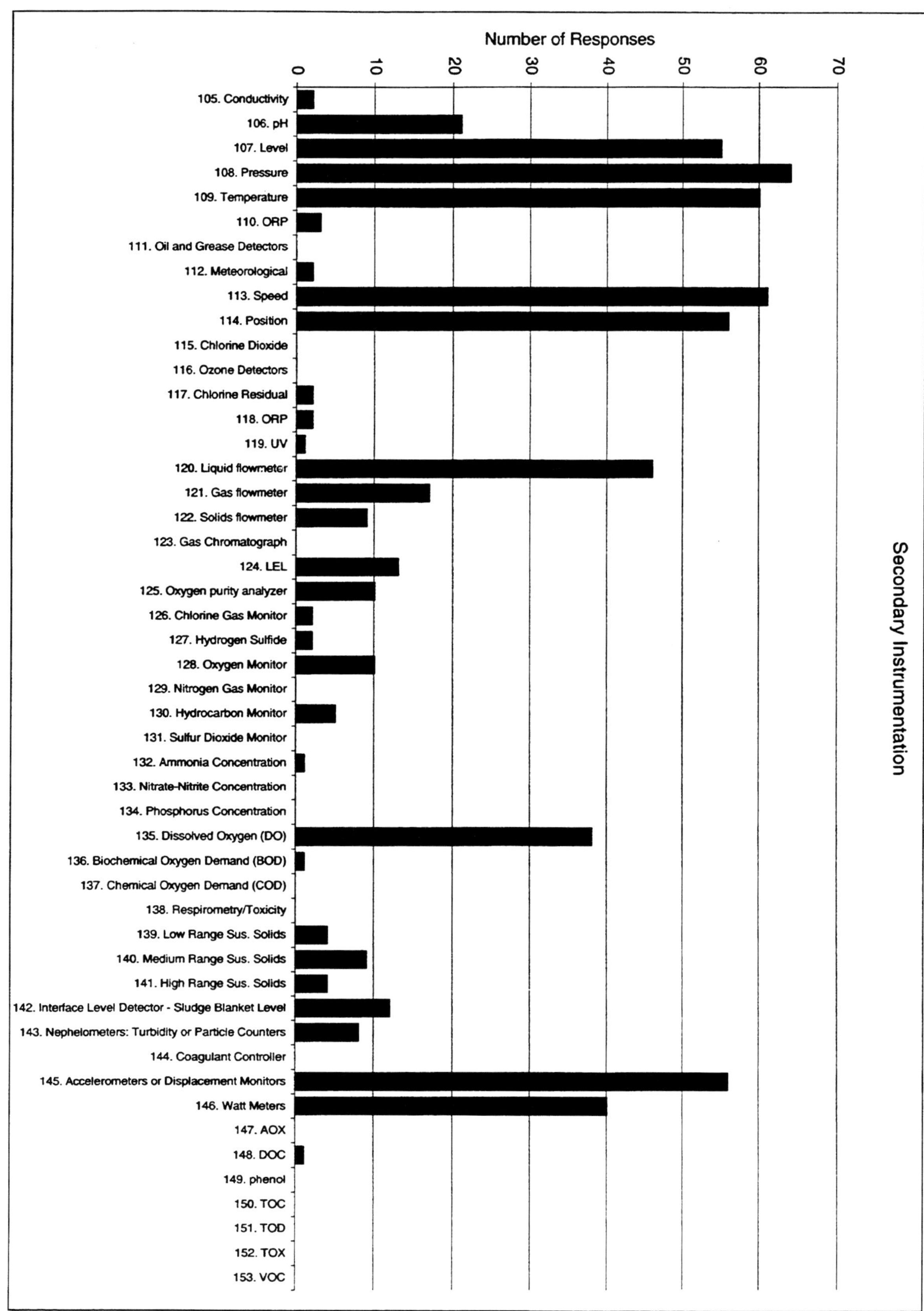

Figure 2-9. Instrumentation Used for Secondary Treatment

For solids processing which includes digestion, thickening, dewatering, and composting, most surveyed treatment facilities shown in Figure 2-10 reported using:

- level,
- pressure,
- temperature,
- speed,
- position,
- liquid, gas and solids flowmeters,
- LEL,
- accelerometers, and
- Watt meters.

High range suspended solids analyzers and sludge blanket level detectors were also important to the operation of several advanced solids treatment systems.

For effluent and disinfection treatment processes, Figure 2-11 shows that many treatment facilities reported using:

- pH,
- level,
- pressure,
- temperature,
- chlorine residual, and
- liquid flowmeter.

Surprisingly, less than 20% of the respondents listed using chlorine gas monitors, which are often required by design criteria for safety purposes.

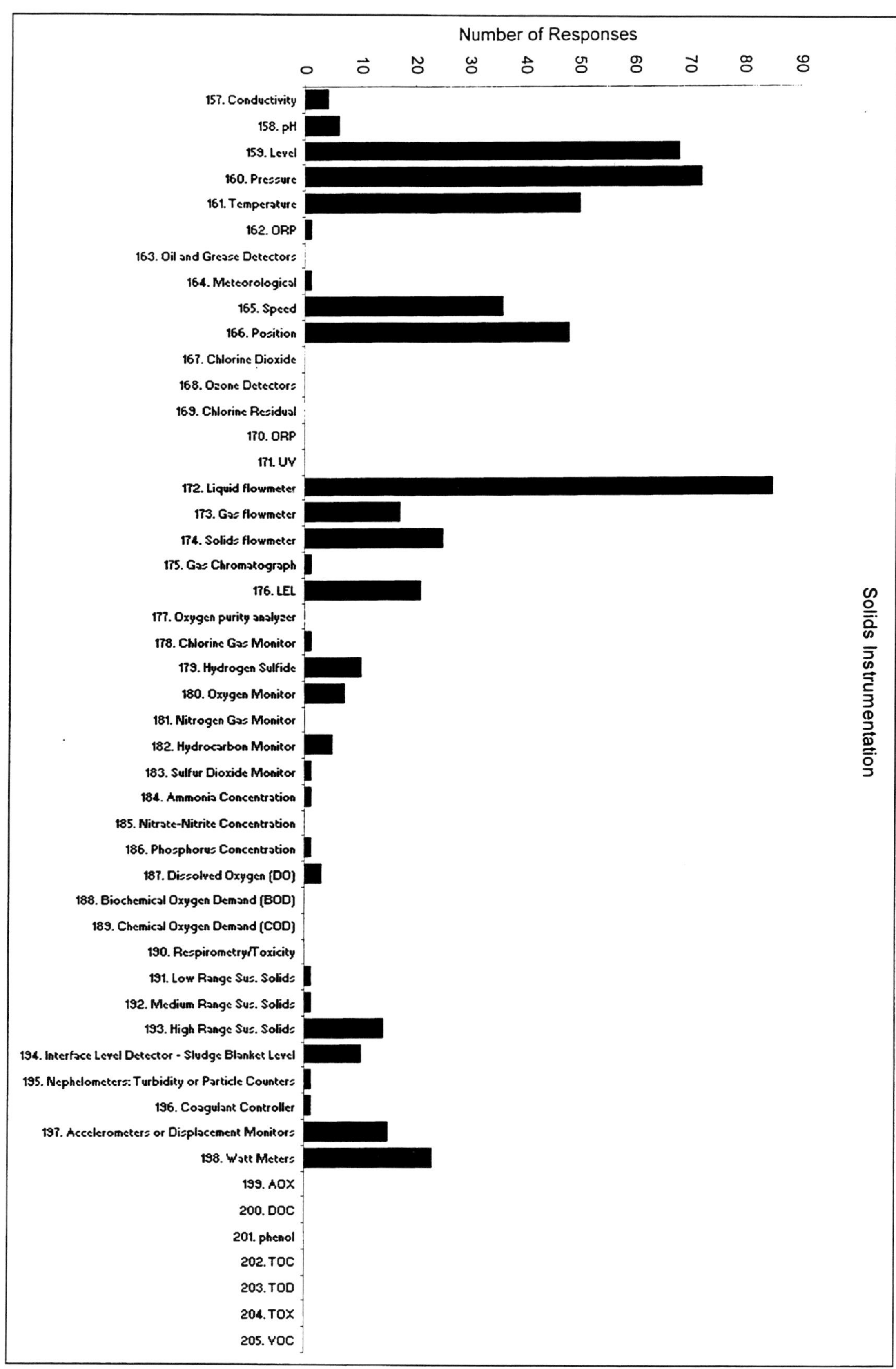

Figure 2-10. Instrumentation Used for Solids Processing

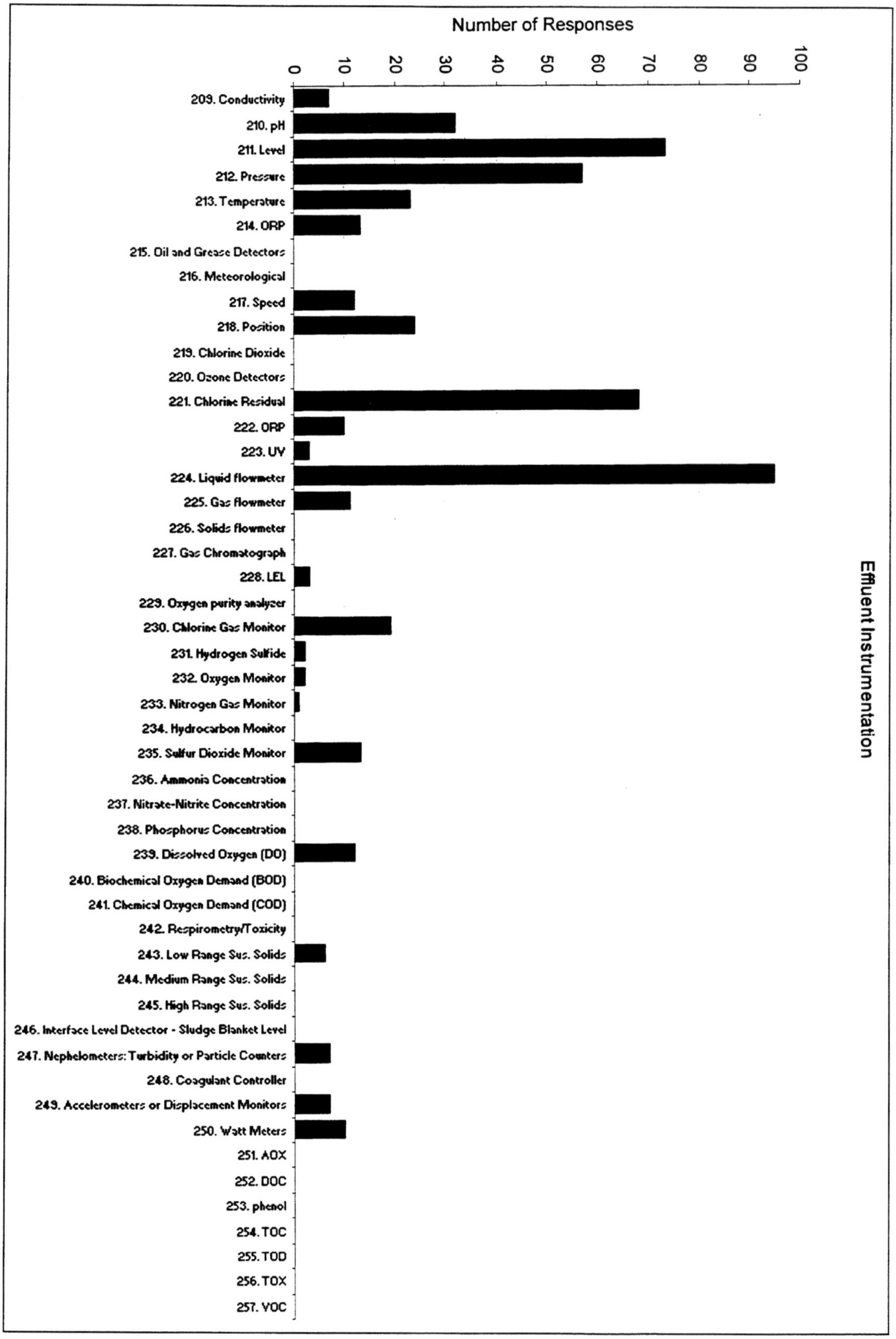

Effluent Instrumentation

Number of Responses

0 10 20 30 40 50 60 70 80 90 100

209. Conductivity
210. pH
211. Level
212. Pressure
213. Temperature
214. ORP
215. Oil and Grease Detectors
216. Meteorological
217. Speed
218. Position
219. Chlorine Dioxide
220. Ozone Detectors
221. Chlorine Residual
222. ORP
223. UV
224. Liquid flowmeter
225. Gas flowmeter
226. Solids flowmeter
227. Gas Chromatograph
228. LEL
229. Oxygen purity analyzer
230. Chlorine Gas Monitor
231. Hydrogen Sulfide
232. Oxygen Monitor
233. Nitrogen Gas Monitor
234. Hydrocarbon Monitor
235. Sulfur Dioxide Monitor
236. Ammonia Concentration
237. Nitrate-Nitrite Concentration
238. Phosphorus Concentration
239. Dissolved Oxygen (DO)
240. Biochemical Oxygen Demand (BOD)
241. Chemical Oxygen Demand (COD)
242. Respirometry/Toxicity
243. Low Range Sus. Solids
244. Medium Range Sus. Solids
245. High Range Sus. Solids
246. Interface Level Detector - Sludge Blanket Level
247. Nephelometers: Turbidity or Particle Counters
248. Coagulant Controller
249. Accelerometers or Displacement Monitors
250. Watt Meters
251. AOX
252. DOC
253. phenol
254. TOC
255. TOD
256. TOX
257. VOC

Figure 2-11. Instrumentation Used for Effluent and Disinfection Processes

2.4 Control Types

The survey quantified the different types of control system technology currently installed throughout North America, as represented by the survey population. To evaluate the different systems used in North American WWTPs, several broad categories of control were defined to describe the general control types available. To qualify how controls are implemented in plants, survey data related to controls used within each plant process area were gathered. The survey was designed to look for trends within each process area in a plant and then within each plant size range. Survey respondents defined the types of systems implemented by process area, defining greater and lesser levels of automation within treatment facilities. Survey responses were geared to gather information about all types of technology used in a facility, so multiple control types were selected for many facilities.

2.4.1 Control Hardware

Control hardware used at treatment facilities covers a wide range of available technology, as shown in Figure 2-12. The specific process area within a treatment facility was analyzed for the type of control hardware.

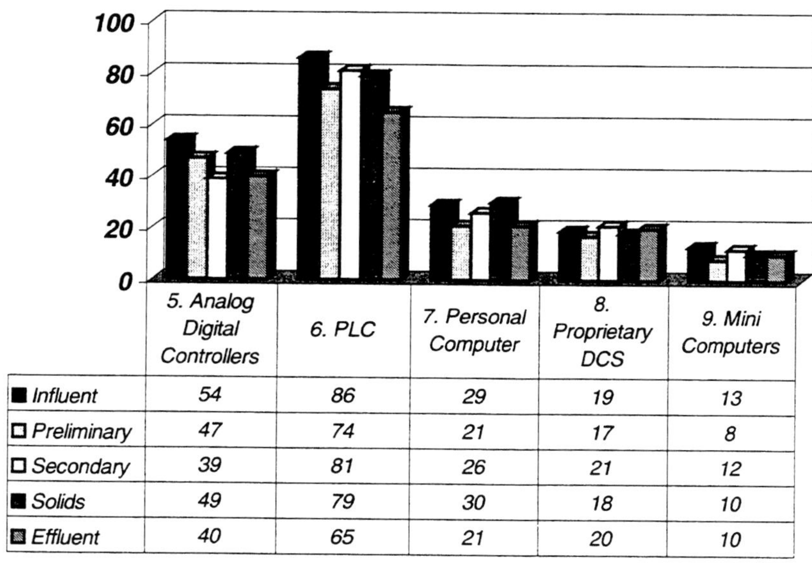

	5. Analog Digital Controllers	6. PLC	7. Personal Computer	8. Proprietary DCS	9. Mini Computers
Influent	54	86	29	19	13
Preliminary	47	74	21	17	8
Secondary	39	81	26	21	12
Solids	49	79	30	18	10
Effluent	40	65	21	20	10

Figure 2-12. Control Hardware Used in Each Process Area

The general classes of hardware are:

- analog/digital controllers or single loop controllers used to control a single control variable with a PID algorithm;
- programmable logic controllers (PLCs) used to control multiple devices, and either stand-alone or networked back to a central location;
- personal computers used to monitor and control facilities;

- proprietary distributed control systems (DCS), including control equipment distributed throughout a facility and networked back to allow monitoring and control from a single location; and
- minicomputers used to process all control configuration based on process conditions monitored through remote I/O processing equipment.

As expected several different types of hardware are installed in each process area. Based on survey results presented in Figure 2-12, the control hardware used most frequently is the PLC. Next most frequently used are analog digital controllers (single loop controllers). Personal computers, proprietary DCS, and mini computers were least often used in facilities.

The variation in application by process area was notable for all types of hardware except proprietary DCS, possibly because DCS are implemented on a plant-wide basis. Influent always registered the greatest number of survey responses for each type of technology, possibly because it is one area that is important to automate for continuous operation. Another possible reason for its relatively higher number of responses relates to the fact that it was the first of five similar forms and may have received the most attention from survey respondents. Secondary treatment facilities were typically the second most heavily automated process area. This is particularly notable because the number of responses dropped off for analog controllers and personal computers, but remained high for PLCs, proprietary DCS, and minicomputers.

When each type of technology is considered based on the size of the facility, some definite trends emerge. Figure 2-13 reflects the percent of facilities of a specific size reporting any use of each technology. The analysis counts any single reported use in any process area as being a positive occurrence within the facility.

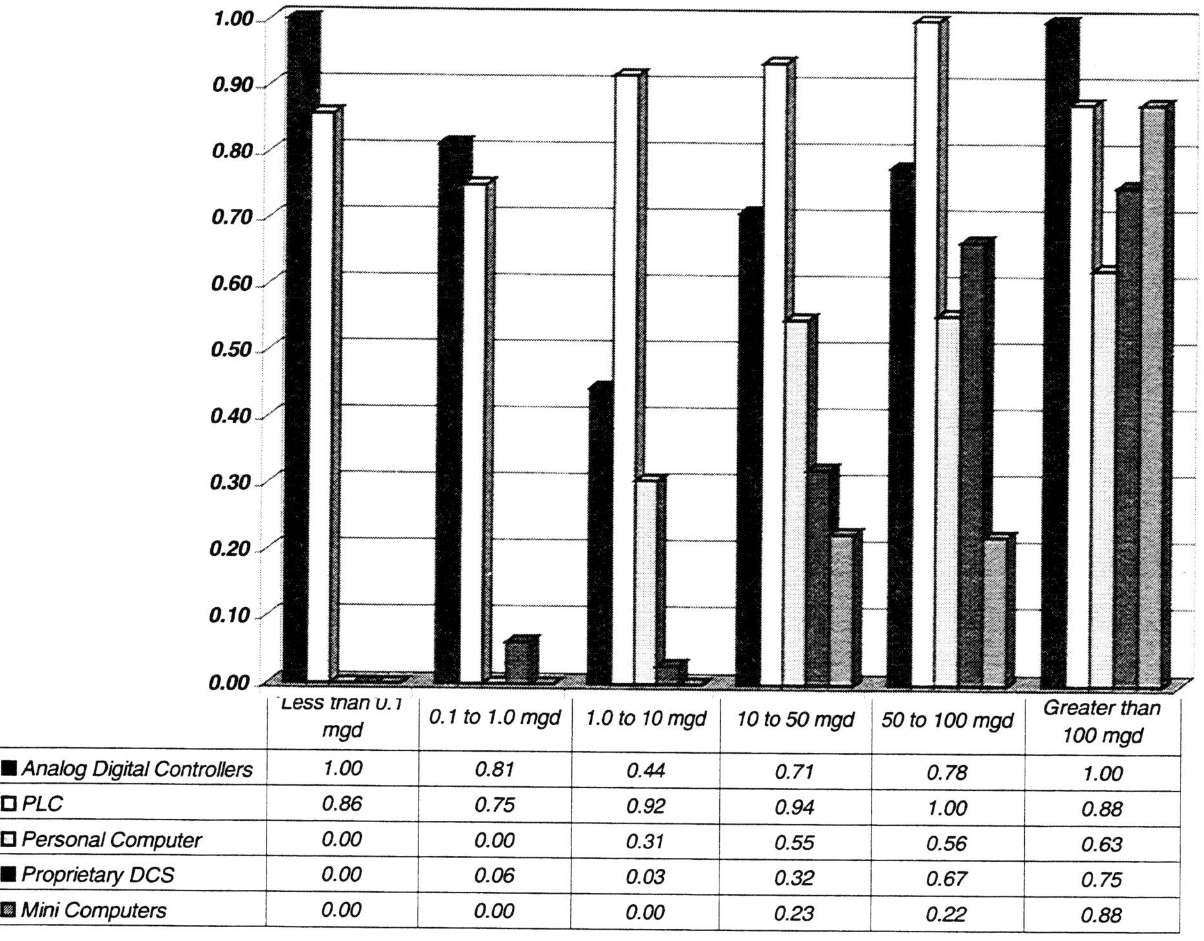

	Less than 0.1 mgd	0.1 to 1.0 mgd	1.0 to 10 mgd	10 to 50 mgd	50 to 100 mgd	Greater than 100 mgd
■ Analog Digital Controllers	1.00	0.81	0.44	0.71	0.78	1.00
□ PLC	0.86	0.75	0.92	0.94	1.00	0.88
□ Personal Computer	0.00	0.00	0.31	0.55	0.56	0.63
■ Proprietary DCS	0.00	0.06	0.03	0.32	0.67	0.75
▩ Mini Computers	0.00	0.00	0.00	0.23	0.22	0.88

Figure 2-13. Control System Use Changes with Size of Plant

As expected, the larger a facility the higher the percent of facilities reporting use of PCs, proprietary DCS, and minicomputers. The use of both PLCs and analog/digital controllers is high at all facilities.

2.4.2 Control Network

Control networks provide the link between field instruments, control processing units, operator workstations, historical data systems, and enterprise applications. Technology for control networks has evolved over the years to include a wide range of network solutions, mostly incompatible with each other. The user community has been demanding compatibility in networks and open systems, allowing them to purchase network solutions from an open market of vendor.

Figure 2-14 indicates that the greatest number of control networks listed are proprietary networks.

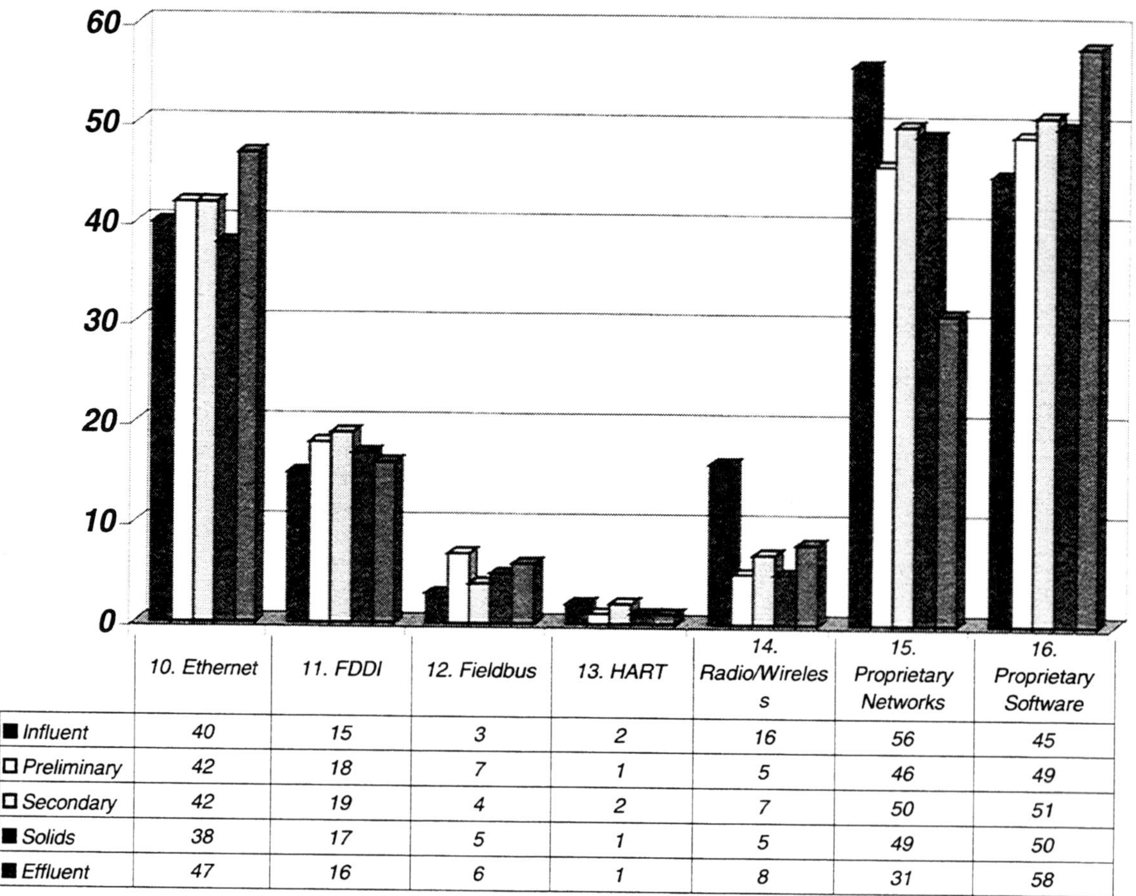

	10. Ethernet	11. FDDI	12. Fieldbus	13. HART	14. Radio/Wireles s	15. Proprietary Networks	16. Proprietary Software
■ Influent	40	15	3	2	16	56	45
□ Preliminary	42	18	7	1	5	46	49
□ Secondary	42	19	4	2	7	50	51
■ Solids	38	17	5	1	5	49	50
■ Effluent	47	16	6	1	8	31	58

Figure 2-14. Control System Network Use in Each Process Area

The number of proprietary networks would be expected, since Ethernet, although now widely available, has only recently been used and recommended for process control systems applied to the wastewater industry. Ethernet is the second most widely reported control network, and a close second to the proprietary networks. Fiber data distributed interface (FDDI) is reported to be used fairly extensively; however, during the site visits, several facilities were identified as using fiber for Ethernet when they had reported FDDI as their control network. For radio/wireless usage, influent showed more than two times the application than at any other process area. The increased use of radio/wireless for influent could be accounted for based on the need to connect remote headworks and other collection facilities with influent treatment. Finally, Fieldbus and HART, two proprietary instrument networks, show very limited use.

When each type of control network is considered based on the size of the facility, some definite trends emerge. Figure 2-15 reflects the percent of facilities of a specific size reporting any use of each type of network. The analysis counts any single reported use in any process area as being a positive occurrence within the facility.

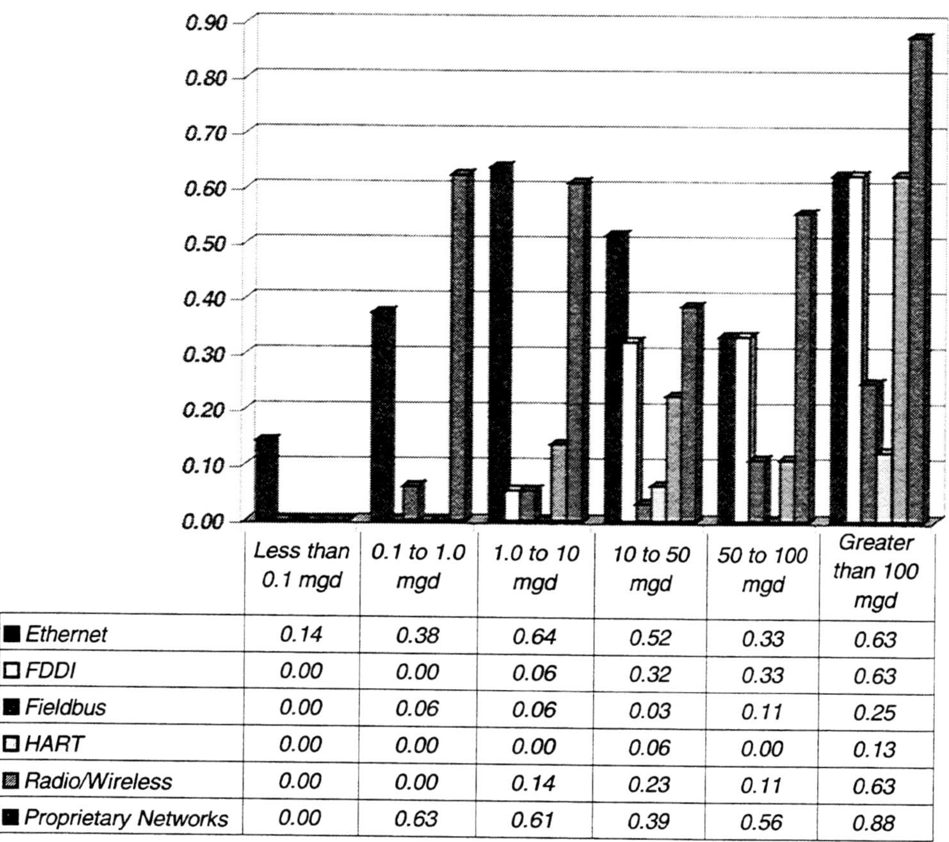

	Less than 0.1 mgd	0.1 to 1.0 mgd	1.0 to 10 mgd	10 to 50 mgd	50 to 100 mgd	Greater than 100 mgd
Ethernet	0.14	0.38	0.64	0.52	0.33	0.63
FDDI	0.00	0.00	0.06	0.32	0.33	0.63
Fieldbus	0.00	0.06	0.06	0.03	0.11	0.25
HART	0.00	0.00	0.00	0.06	0.00	0.13
Radio/Wireless	0.00	0.00	0.14	0.23	0.11	0.63
Proprietary Networks	0.00	0.63	0.61	0.39	0.56	0.88

Figure 2-15. Control System Network Varies Depending on Plant Size

Figure 2-15 indicates that facilities with flows less than 1 mgd have relatively little network usage. Proprietary networks are used most frequently, with all size groupings of facilities consistently showing almost half, or more, using proprietary networks. Larger facilities appear to have a greater percentage of control networks implemented, and more different types of control networks.

2.4.3 Control Software

Control software encompasses programs and applications written to perform a wide range of applications including process control and user interface. As control systems have developed, so has the software that provides a means for users to customize and configure the control systems.

Owners take many different approaches to implementing their systems, ranging from buying packaged systems, to hiring a vendor, system integrator, or engineering firm to develop custom applications, to purchasing the system and developing the control applications and operator interface in house, using employees.

Figure 2-16 indicates that the percentage of each type of software in use reported by owners correlates to the number of systems installed in each process area.

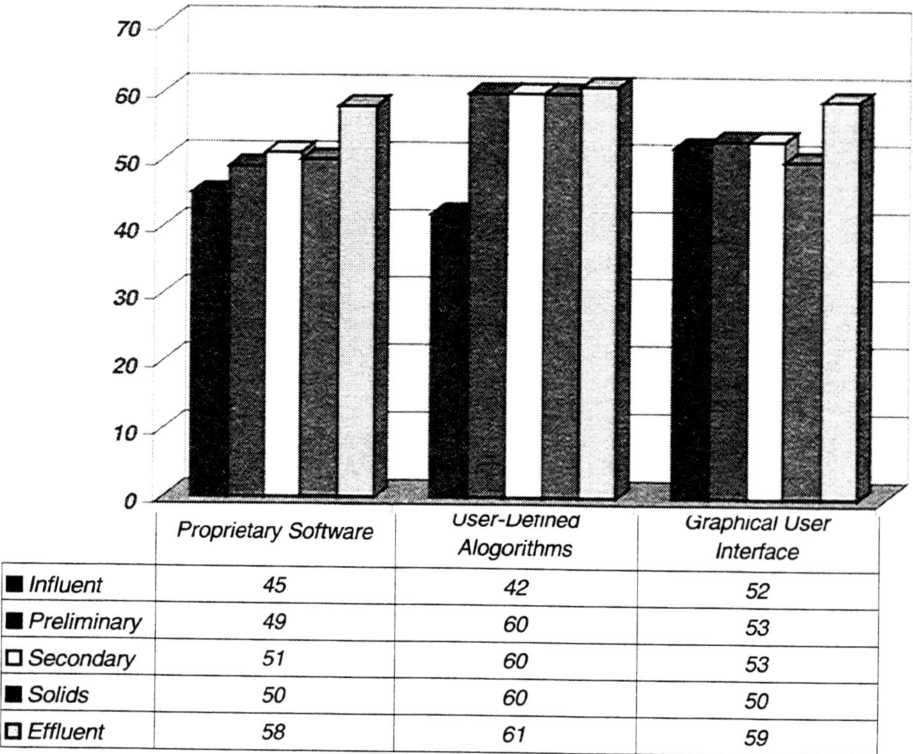

	Proprietary Software	User-Defined Alogorithms	Graphical User Interface
■ Influent	45	42	52
■ Preliminary	49	60	53
□ Secondary	51	60	53
■ Solids	50	60	50
□ Effluent	58	61	59

Figure 2-16. Control Software Depending on Process Area

When each type of control software is considered based on the size of the facility, some definite trends emerge. Figure 2-17 reflects the percent of facilities of a specific size reporting any use of each type of control software. The analysis counts any single reported use in any process area as being a positive occurrence within the facility.

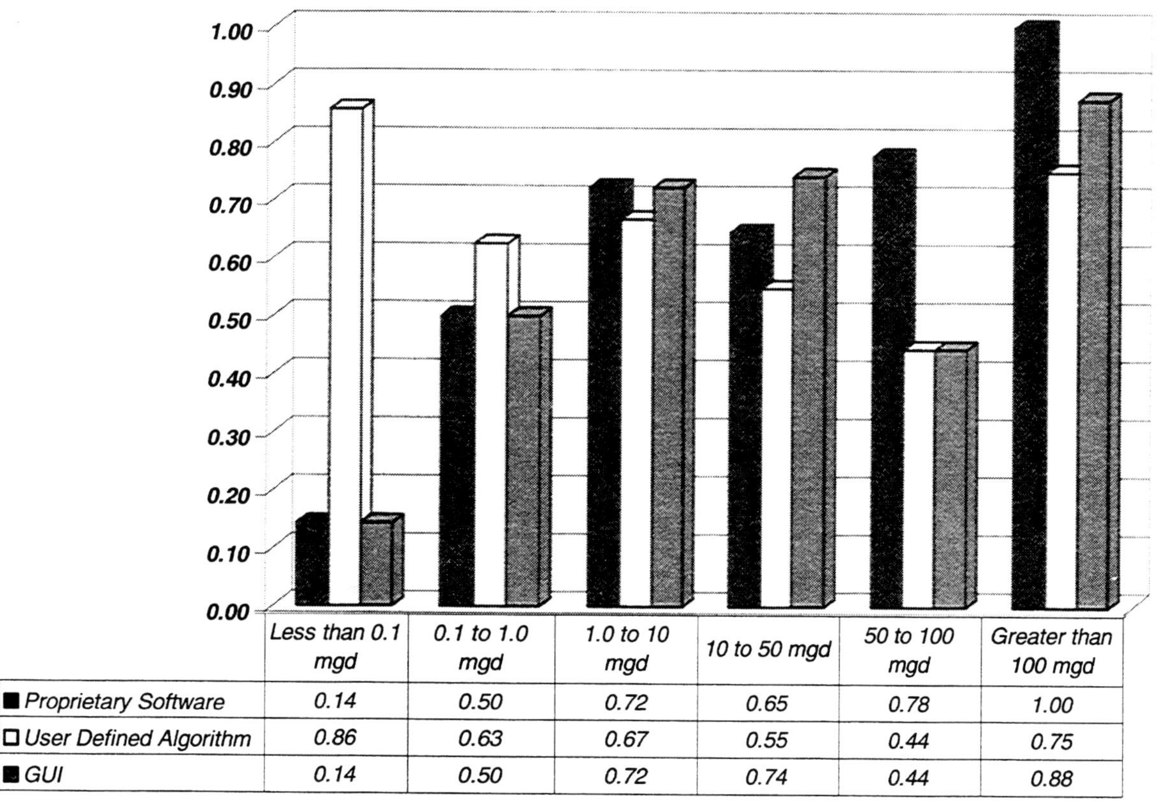

	Less than 0.1 mgd	0.1 to 1.0 mgd	1.0 to 10 mgd	10 to 50 mgd	50 to 100 mgd	Greater than 100 mgd
■ Proprietary Software	0.14	0.50	0.72	0.65	0.78	1.00
□ User Defined Algorithm	0.86	0.63	0.67	0.55	0.44	0.75
■ GUI	0.14	0.50	0.72	0.74	0.44	0.88

Figure 2-17. Control Software Depending on Plant Size

Figure 2-17 indicates that the percent reporting use of proprietary software increased with plant size. The percent of facilities reporting using user-defined algorithms did not show any specific trend with plant size, indicating that there must be other reasons for using user-defined algorithms. Finally, graphical user interface (GUI) use generally increased with increasing plant size.

2.4.4 Age of System

The survey requested information related to the last system upgrade. The age of the system may have affected some of the technology-related questions.

Figure 2-18a-e indicates that most respondents who reported recent upgrades upgraded in the past four years. With approximately half the respondents reporting any date for the most recent upgrade, the staff completing the survey may not have known the date and so left the information blank. Another possible explanation for the large number of upgrades in the past four years relates to Y2K preparedness.

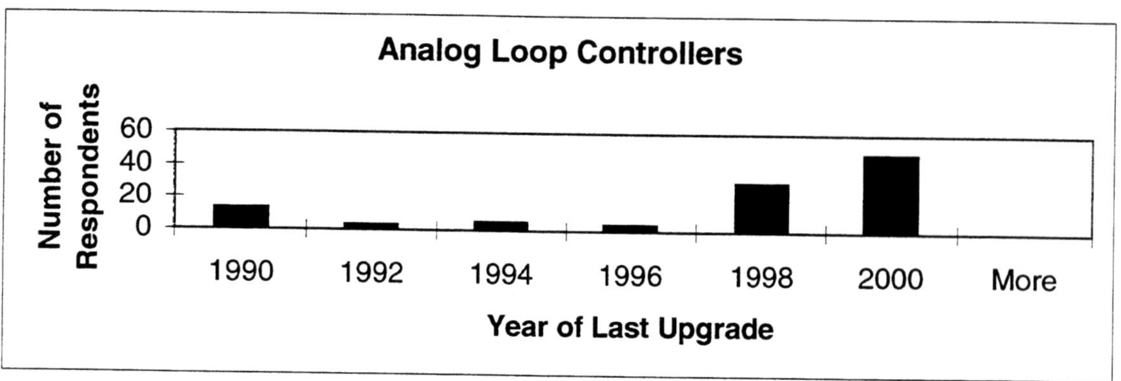

Figure 2-18a. Year of Last Upgrade for Analog Loop Controllers

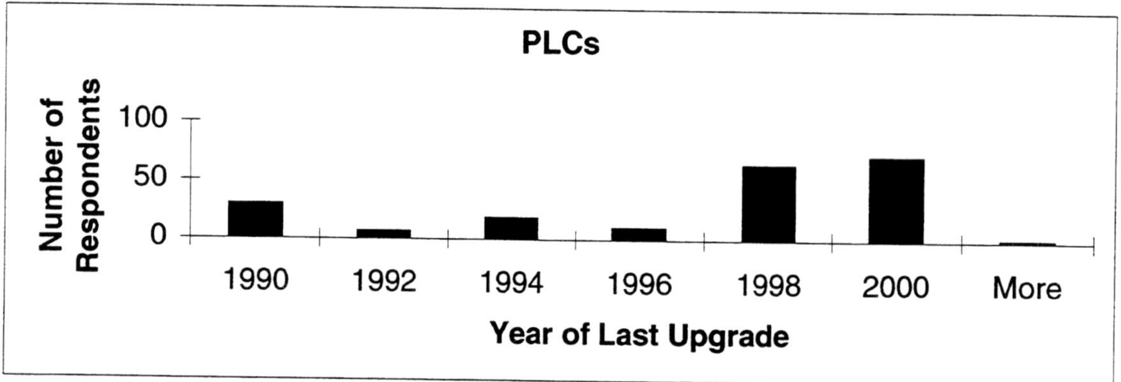

Figure 2-18b. Year of Last Upgrade for PLCs

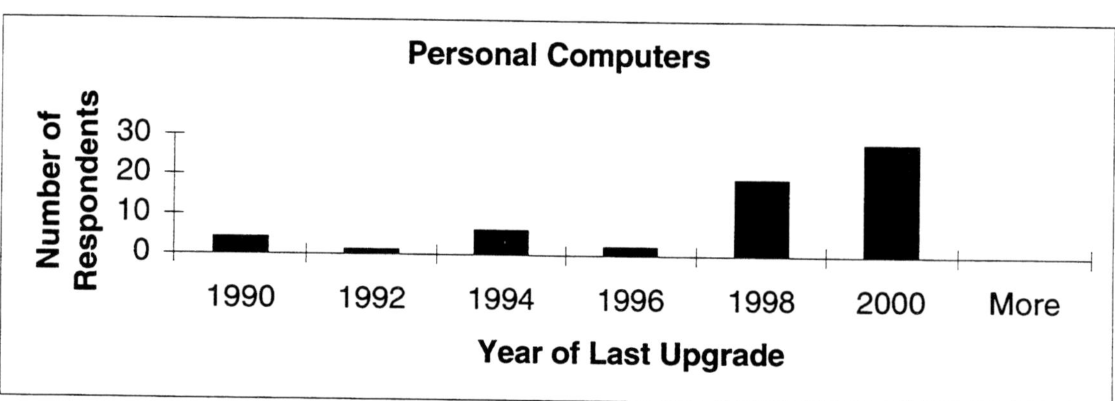

Figure 2-18c. Year of Last Upgrade for Personal Computers

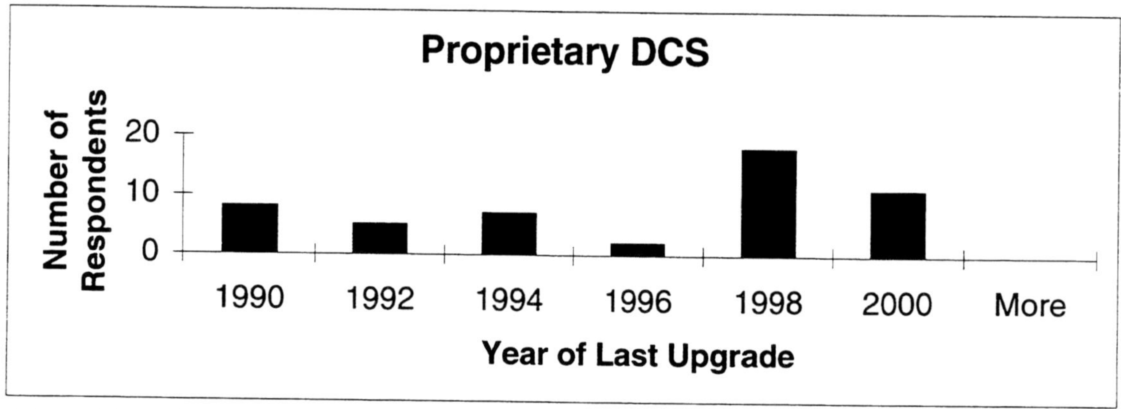

Figure 2-18d. Year of Last Upgrade for Proprietary DCS

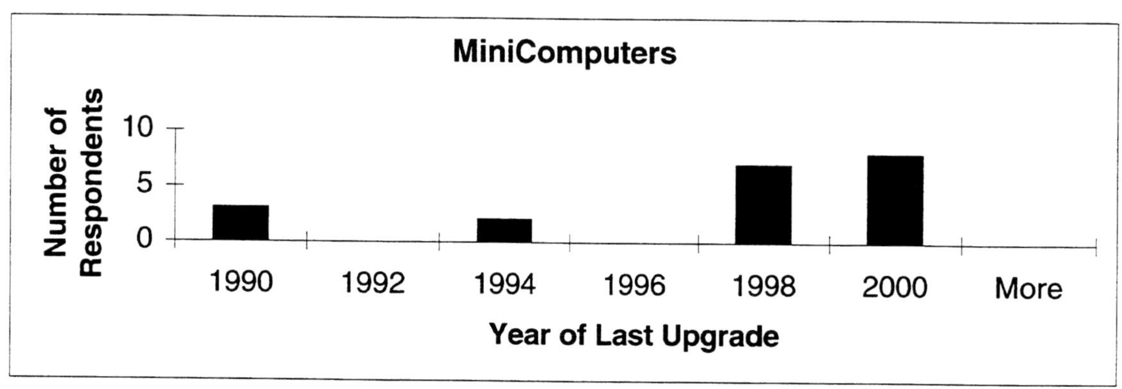

Figure 2-18e. Year of Last Upgrade for Minicomputers

Similar to control system hardware, the control system networks were also upgraded during the past few years, as shown in Figure 2-19a-d.

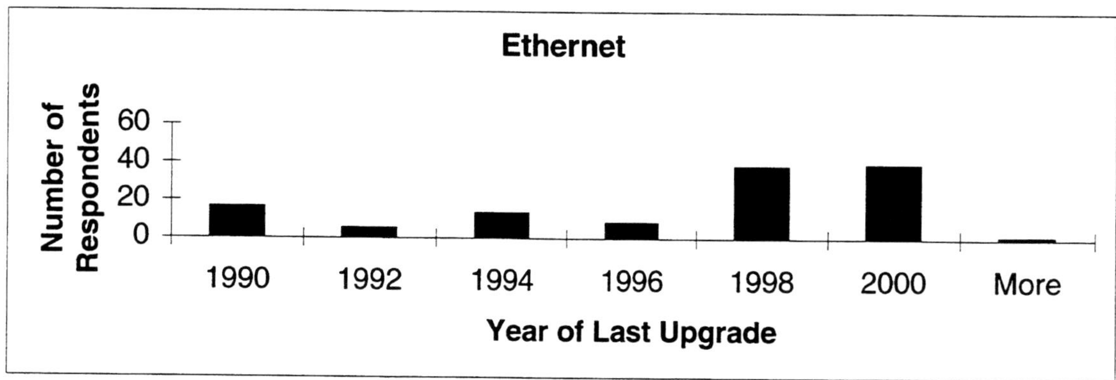

Figure 2-19a. Year of Last Upgrade for Ethernet

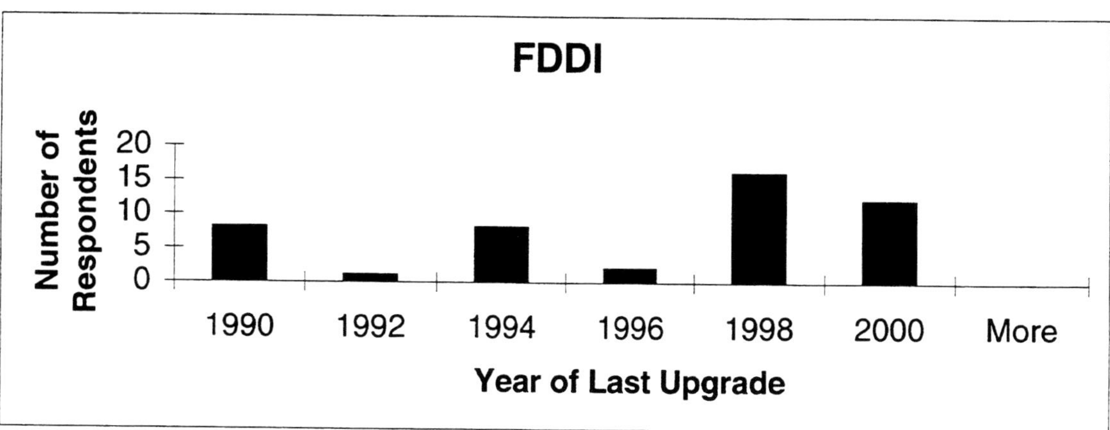

Figure 2-19b. Year of Last Upgrade for FDDI

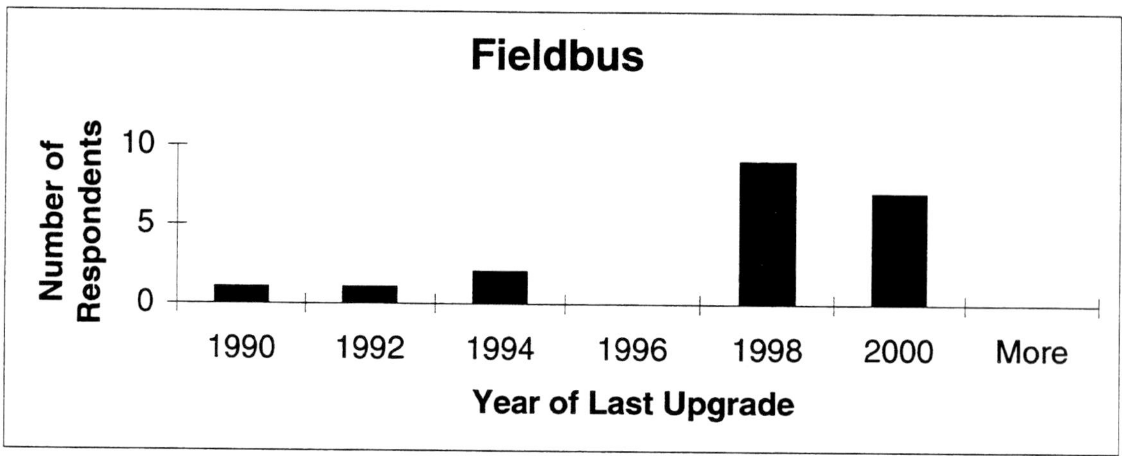

Figure 2-19c. Year of Last Upgrade for Fieldbus

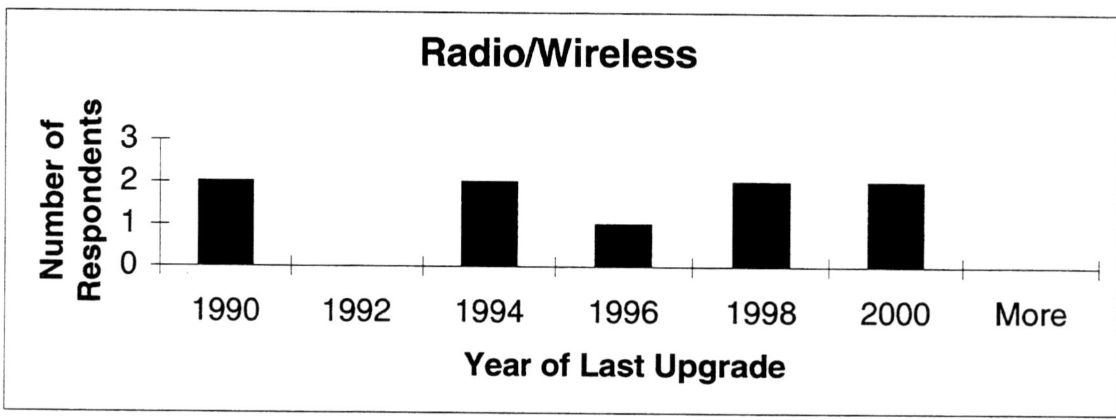

Figure 2-19d. Year of Last Upgrade for Radio/Wireless

Chapter 3.0

FIELD SURVEY

3.1 Introduction and Background

The proposed several different methods for obtaining information from municipal and industrial wastewater utilities. Since the initial I&C survey was structured to allow easy comparison among facilities, the survey had limited capacity for gathering subjective information about facilities. However, site-specific details were essential for the analysis and reporting to make the project meaningful. Therefore, the project approach called for field surveys to gather detailed information about individual facilities.

The field survey provided a mechanism to gather detail from specific facilities through one-on-one interviews with field staff. The field surveys also provided an opportunity to discuss control specifics and to delve into the details of sites' I&C applications. Through dialogue structured around the field survey form, site staff could discuss the details of their I&C applications, ranging from use of sensors, to control applications, to challenges and resulting solutions, to specific difficult issues encountered at the facility.

3.1.1 Criteria for Selecting Sites for Field Surveys

The sites selected for field survey were picked based on several different criteria. Most important was selecting a range of facilities, so that different technologies and applications were represented in the sample.

3.1.1.1 Plant Control System Size Compared to Plant Size

Plant capacity was compared to I/O count. Plants reporting relatively high I/O quantity to mgd were considered candidates for the field survey on the basis that the high I/O count reflected a higher degree of sophistication for the process control system. A high I/O count may indicate that the controls may have been used to implement some advanced or sophisticated control applications.

3.1.1.2 Instrument and Sensor Use

The types of instruments identified in use at a facility were considered to be an indicator of the level of sophistication of controls at the facility. Facilities indicating relatively uncommon instruments in use like ammonia analyzers, ORP and Cl_2 analyzers, and solids analyzers were considered for field survey, especially when team staff had prior knowledge of sophisticated controls.

3.1.1.3 Process Applications

Sophisticated process applications were used to identify facilities that should undergo field survey. Facilities with unusual or advanced control applications were identified primarily based on specific sensor use identified in the survey, and on the project team members' prior knowledge and experience. Several facilities that had not completed the initial I&C survey were added to the list of candidate locations because of specific control applications that were known to have been implemented at the facilities.

3.1.1.4 Plant Running Unattended

Facilities reporting unattended operation, especially relatively large facilities, were considered to be excellent candidates for field survey.

3.1.1.5 Cost-Benefit Data Justifying I&C Projects Available

Almost no facilities indicated that they cost justify expenditure on process control by showing any tangible benefits. Where facilities stated that they did have cost-benefit data, a field survey was considered, in order to analyze the effectiveness and presentation of the cost data.

3.1.1.6 Industrial Facilities

Very few industrial facilities responded to the written survey; however, the Project Advisory Committee made recommendations for site visits to wastewater treatment facilities associated with paper and chemical plants. The four facilities visited demonstrated a range of automation from manual operations only to a highly automated facility. The industrial WWTP facilities were part of a larger chemical or pulp and paper operation but typically had staff dedicated to the operation of the WWTP, independent of the rest of the plant or mill.

Using the criteria listed above, 38 facilities and utilities were chosen as candidates for a site survey.

Another factor taken into consideration was geography. Facilities in close proximity to the people doing surveys were obvious candidates for survey, since travel/time costs would be reduced. Several other site visits were scheduled to combine facilities located close to each other. Once candidate facilities for field survey were identified, facilities were grouped based on location, and site visits planned to encompass as many facilities as possible in a short amount of time.

3.2 Demographics

The following table identifies the facilities identified and the reasons for visiting them. Duffin Creek was included based on team knowledge, and the two Louisville plants were "targets of opportunity," since an advisory team member is a senior manager at the utility.

Table 3-1. Summary of Facilities Visited During Field Survey

Plant Name	*High I/O to MGD Ratio*	*Sensor Use*	*Process Applications*	*Unattended Operations*	*Cost Benefit Data*	*Industrial Facility*
MWRA – Deer Island Treatment Plant	High I/O count		Large plant size, H2S corrosion control			
Houston – Sims South			Aeration control			
Clark County (CCSD)			SRT, Alkalinity Control			
San Jose CA			SRT control, primary treatment			
Orange County, FL – NW WRF	High I/O Count (1-10 MGD)		Alkalinity, Disinfection		yes (for all columns)	
Houston 69th Street			Gravity thickening control Dechlorination control			
Madison WI			UV disinfection			
Collingwood			Phosphorous control			
Roseville, CA			Overall control system			
Houston – Southwest (see row 2 above)				Unattended operations		
Henderson NV	High I/O (10 – 50 mgd)					
Chicago Stickney	High I/O (>100 mgd)					
United Defense L.P.						WTP for industrial waste
Blue Ridge Paper		On-line respirometer				WWTP for paper mill
Anonymous Paper Plant			Reported high degree of automation			WWTP for paper mill (produces tissue)
Eastman Chemical Company Kingsport, TN			Reported high degree of automation			WWTP for chemical plant
Boise Cascade International Falls, MN						WWTP for Paper mill
City of Las Vegas, NV					yes	
DuPont Cooper River						WWTP for fiber production
DuPont Dow Elastomers, Louisville, KY						WWTP for fiber production
Duffin Creek WWTP, Ont.						
North Shore Sanitation District, IL				Unattended operations of BFP	Cost-benefit data	
San Francisco, CA					Cost data	
Morris Foreman WWTP, Louisville, KY						
Cedar Creek WWTP, Louisville, KY						

3.3 Analysis of Control Strategies Identified During Field Survey

The results of the field surveys are covered in detail in Chapter 4. The following section summarizes the types of control strategies implemented in facilities. The charts are histograms based on the number of reported occurrences of each type of control strategy. The list of strategies was initially developed based on the WEF Special Publication *Automated Process Control Strategies* but has been modified based on actual findings.

3.3.1 Pumping

Facilities reported a wide range of pumping options, based on different configurations of variable speed and fixed speed pumps. Most facilities reported manual operation of influent pumping controls, including manual start/stop and speed adjustment. Several facilities also reported automated pump sequencing to equalize flow (see Figure 3-1).

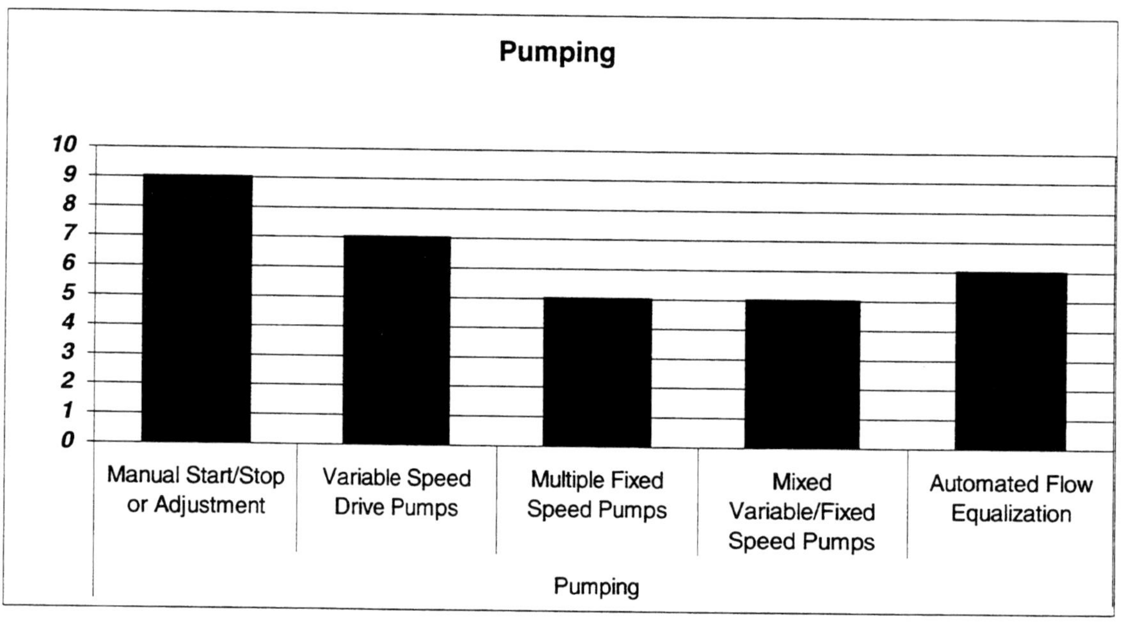

Figure 3-1. Types of Pumping Control Strategies

Three facilities reported both manual and automated control; the rest of the facilities reported one or the other method exclusively. In one of the facilities reporting both manual and automated control, MWRA's Deer Island, the 2,600 kW (3500 hp) variable speed pumps must be manually started and stopped by operators. Once running, pump speed is automatically controlled to match pumped flow to the flow in collection system. When the control strategy identifies pumps operating outside optimum speeds, the strategy notifies the operator to manually start or stop a pump.

Some facilities identified incorrectly sized pumps as a limitation to automated operations. Fixed speed pumps that were not sized correctly to handle the variability in normal plant flows could not be controlled through sequential control strategies. Plant staff monitored flow, or an inlet wet well or sewer level, and used that as the basis to manually initiate plant feed.

3.3.2 Preliminary and Primary Treatment

The main processes identified for automation in preliminary and primary treatment include bar screens, grit removal, and primary pumping (Figure 3-2). Facilities reported nearly equal use of elapsed time and differential level control of bar screen operation. Three of the facilities reported using both manual and some other automated control strategy.

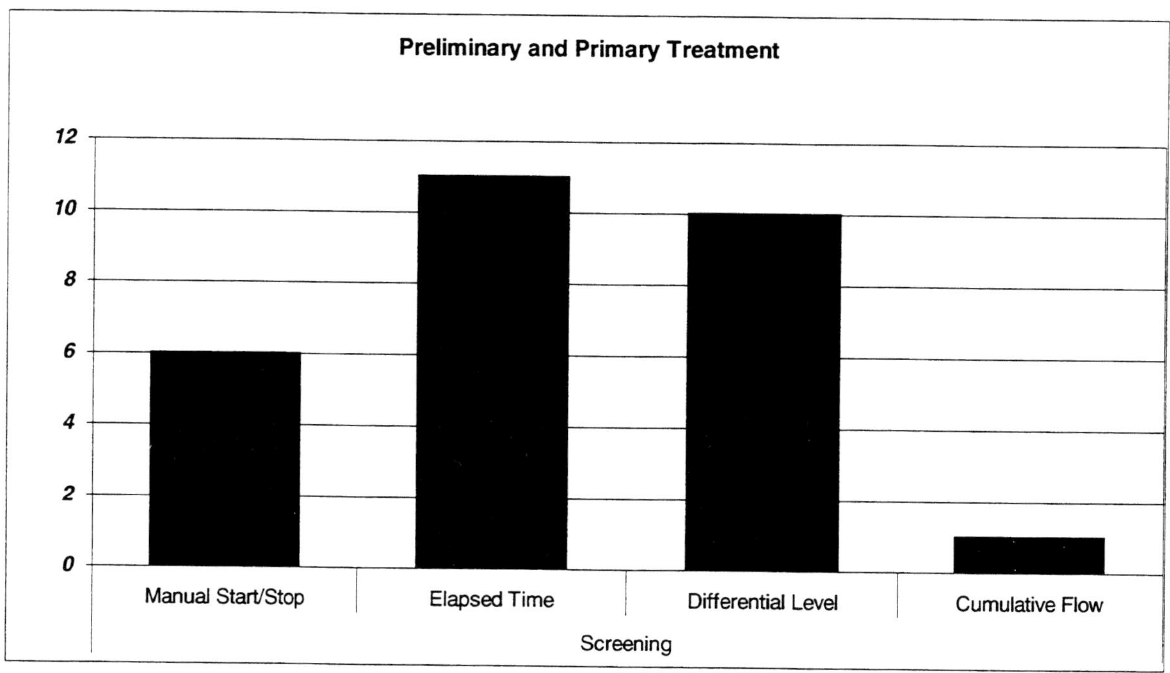

Figure 3-2. Types of Bar Screening Control Strategies

Facilities reported using constant flow to control grit removal, implying that, for the sample group, process construction, rather than process control, provides the most frequently used grit removal strategy (Figure 3-3).

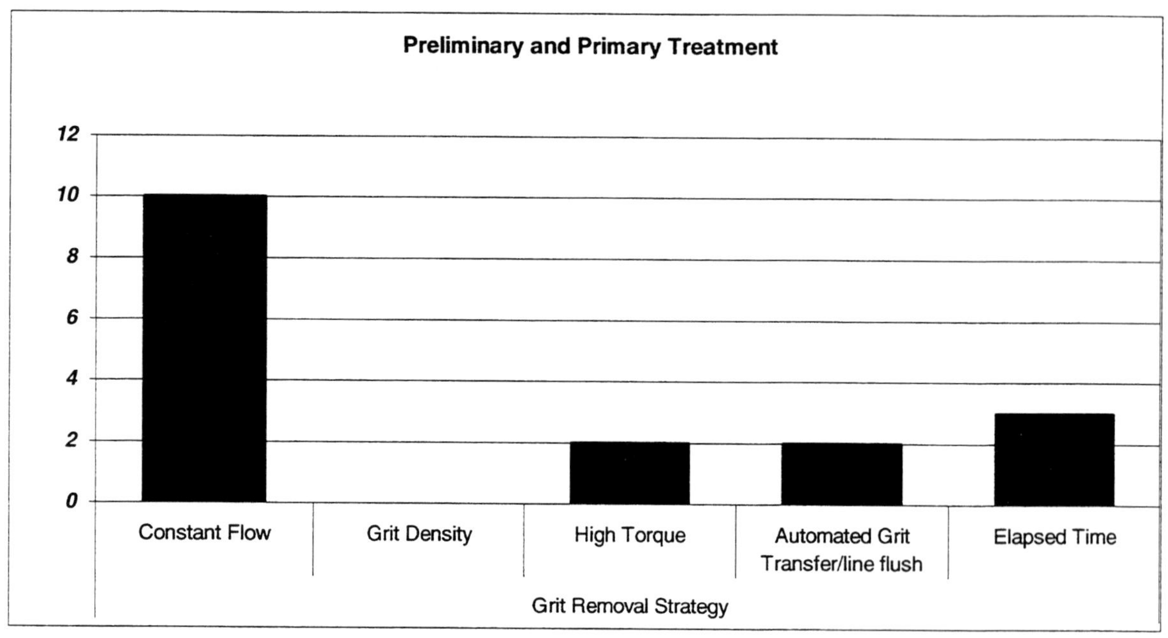

Figure 3-3. Types of Grit Removal Control Strategies

Facilities in the sample group reported using timer-based primary sludge pumping control more than any other alternative (Figure 3-4). Only one plant visited used instrumentation-based primary sludge pumping strategies, including sludge density or depth monitors.

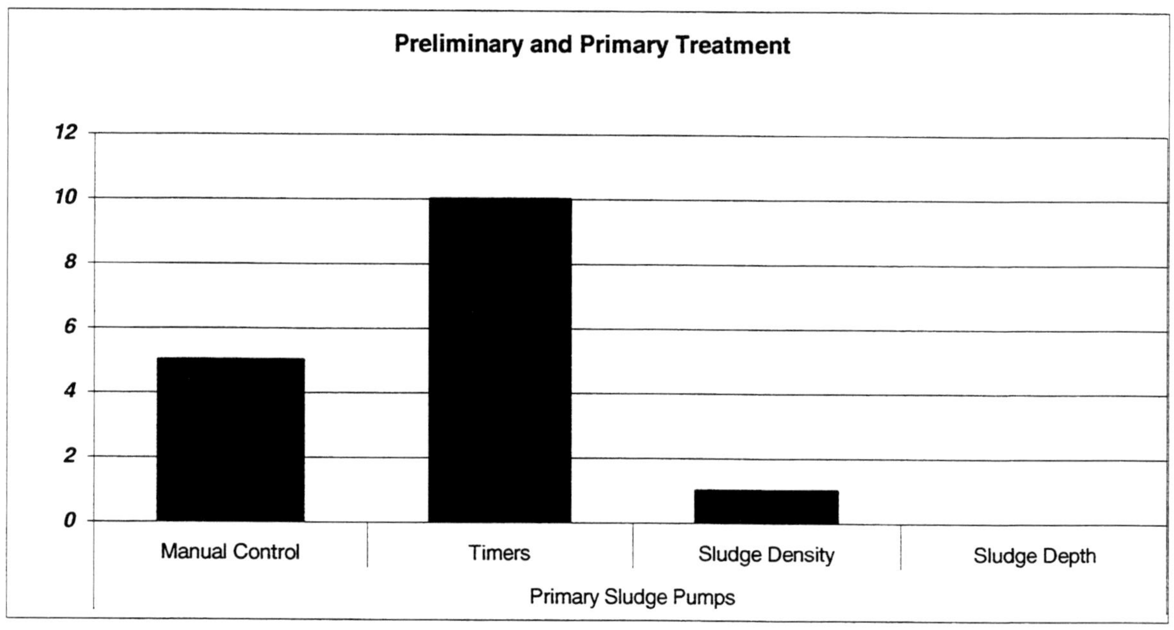

Figure 3-4. Types of Primary Sludge Pumping Control Strategies

3.3.3 Secondary Treatment

The main strategies identified for secondary treatment included blower control, DO control, sludge wasting, and return sludge control. The following charts reflect the frequency of reported occurrence of specific control strategies among the sample group of utilities interviewed during the field survey. As shown in Figure 3-5, the sample group reported automating blowers to control header pressure as the most frequently used blower control strategy.

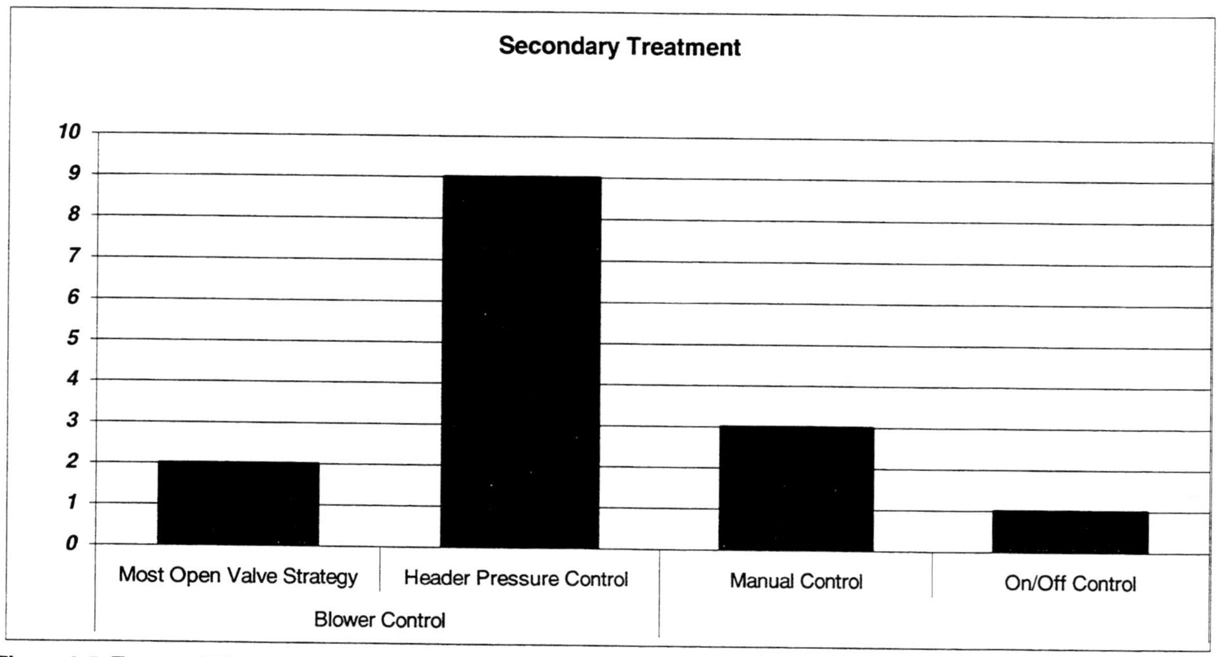

Figure 3-5. Types of Blower Control Strategies

Two facilities reported using the most open valve strategy (see Section 4.6.1). Facilities reporting manual control of blowers reported no other type of automated control strategy in use in addition to the manual control.

As seen in Figure 3-6, facilities reported compound DO/air flow loops to control dissolved oxygen in the secondary process most frequently. Control based only on DO concentration was the next most frequently used strategy identified.

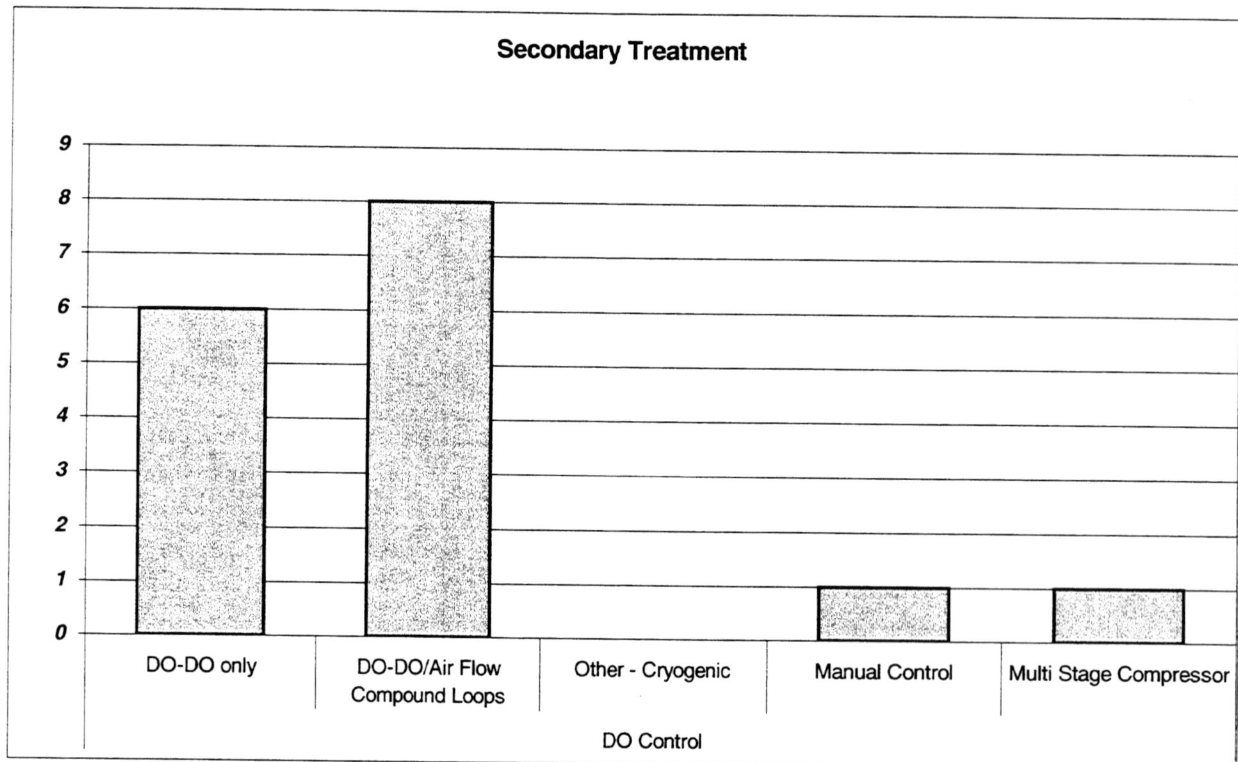

Figure 3-6. Types of Dissolved Oxygen Control Strategies

Facilities in the sample group reported using constant flow rate sludge wasting and sludge age (SRT) control equally frequently (Figure 3-7). Constant flow rate control implies that the wasting rate is set based on lab or other decisions and the flow rate set to match the calculated rate. Several approaches to SRT control were found, some of which are discussed in more detail in Chapter 4 of this report. One facility described inoperative equipment as the reason for continuous wasting.

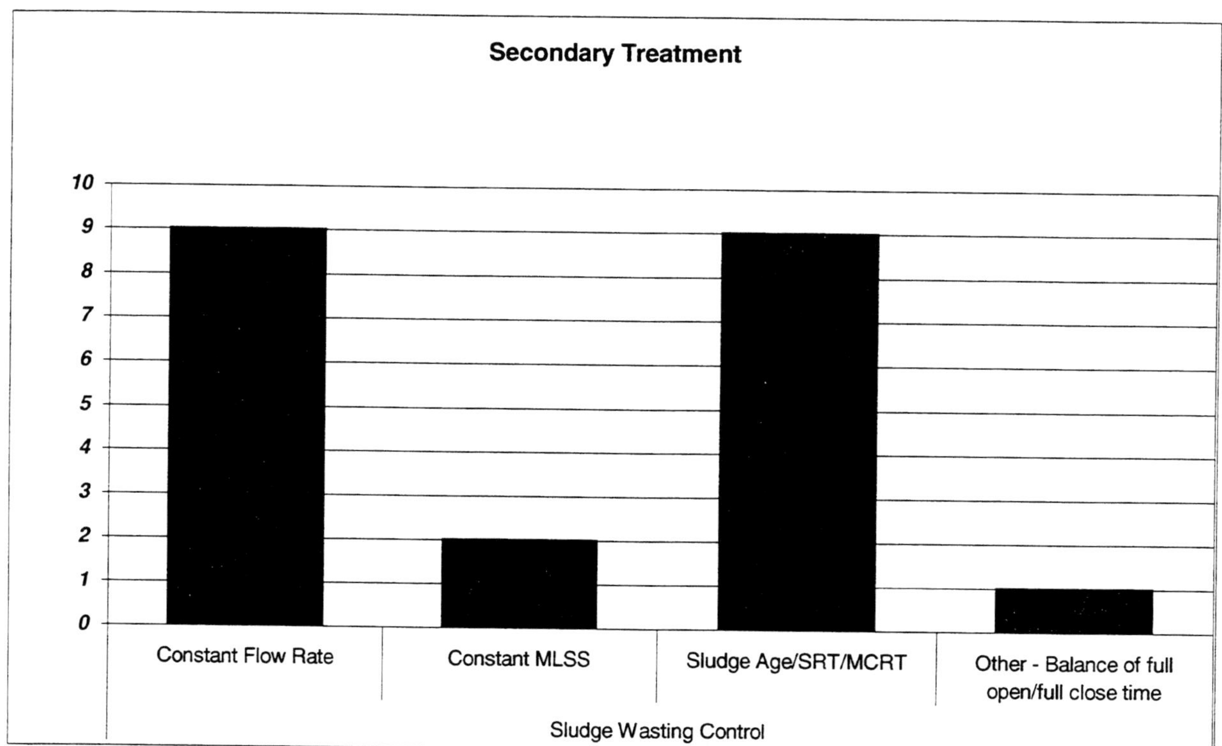

Figure 3-7. Types of Sludge Wasting Control Strategies

As seen in Figure 3-8, facilities in the sample group reported using constant flow rate return sludge control more than any other strategy. This result implies that the return rate is set based on lab or other decisions and the flow rate set to match the calculated rate. Previous research has demonstrated that as long as the return rate is not too low or too high, it has little affect on the operation of the activated sludge process.

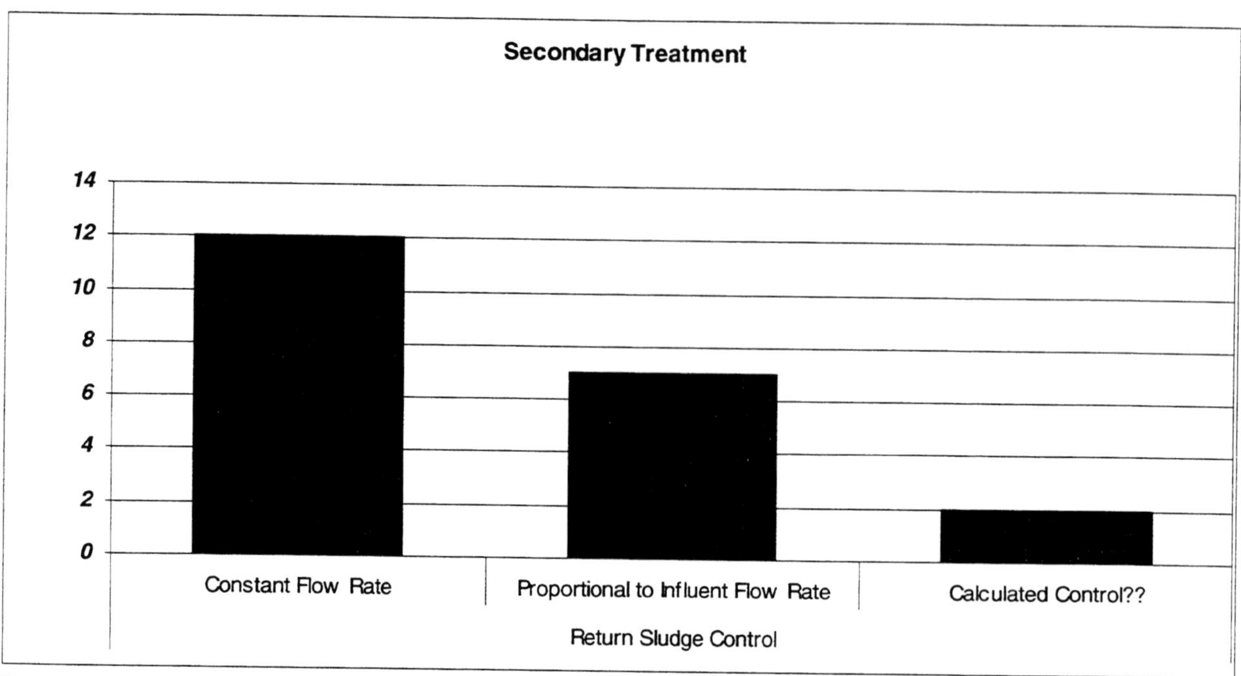

Figure 3-8. Types of Return Sludge Control

One industrial facility, Blue Ridge Paper, reported using on-line respirometers for automatic adjustment of aeration control. This plant is largely operated in manual mode without any automated strategies. The plant staff installed two on-line units, one on the primary effluent and the other on the MLSS flow. The theory associated with respirometer use was to calculate stage of treatment, and to adjust the oxygen or airflow accordingly. The distinct phases of aeration include an initial lag phase, followed by a period of maximum respiration, followed by a period of declining respiration, and finally a period of endogenous rate of respiration. According to the respirometer supplier, the phases can be detected by plotting time vs. O_2 used. Each phase, as depicted on a graph, can be used to characterize the organisms present in the wastewater.

The plant operated at or near maximum capacity at the time the units were installed. Once installed, the units operated to identify when the plant was overloaded, and additional aeration capacity was required. When the plant overflowed, the units did not operate well, due to foaming. An alternate mode of detecting spills was required, including conductivity monitors in the sewers.

The respirometers required between ½ and 4 hours per day of maintenance, generally associated with cleaning the 200-yard-long insulated sample lines. Once the plant flows reduced below plant capacity due to unrelated process changes, the respirometers were abandoned. At that time, they were severely corroded due to poor quality instrument air.

All four industrial facilities involved in the field survey reported using step feed control as part of the secondary treatment process. Only two of the municipal facilities reported using step feed, indicating that either the characteristic of the industrial wastewater is such that step feed control provides a distinct advantage as a treatment option, or that municipal facilities may not be taking advantage of a treatment option available to them.

All four industrial facilities reported sludge wasting at a constant rate, calculated using some variation on sludge age/SRT/ or MCRT.

3.3.4 Disinfection

The main strategies identified for disinfection include chlorination and dechlorination control and UV control.

As seen in Figure 3-9, the most frequently used chlorination strategy defined was flow proportional control; however, almost as many facilities reported using either ORP or compound loop control. A compound loop control strategy for disinfection combines a residual control loop whose output is fed into a flow proportioning control loop.

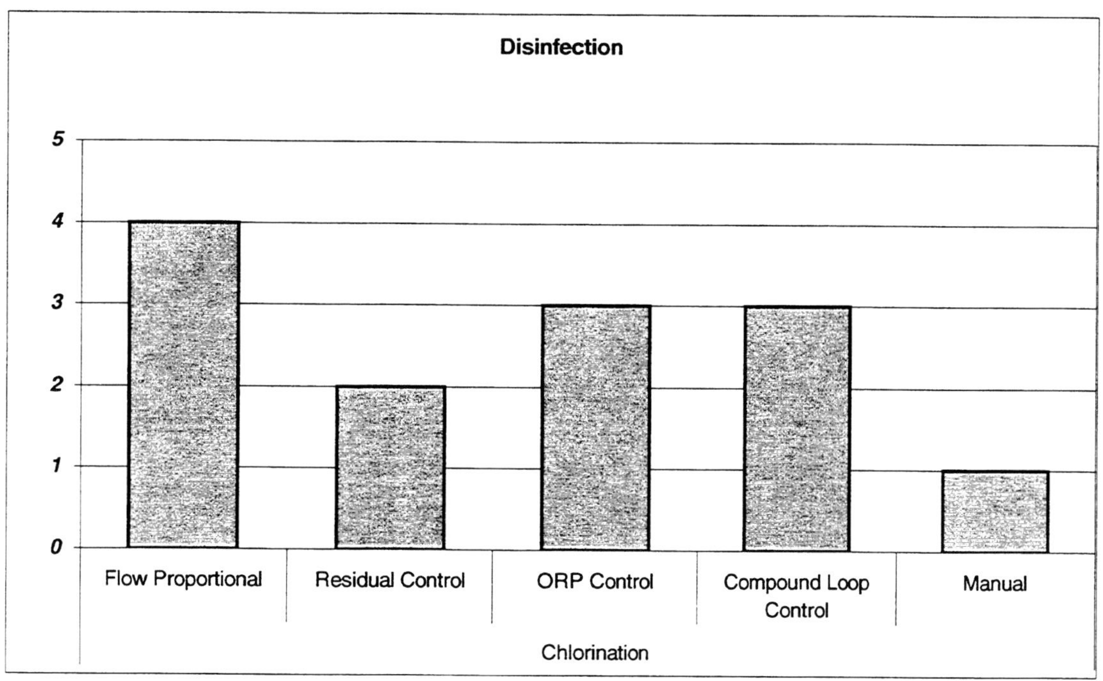

Figure 3-9. Types of Chlorination Control

Most facilities reported using compound loop control for dechlorination control, although many facilities reported using ORP control (see Figure 3-10).

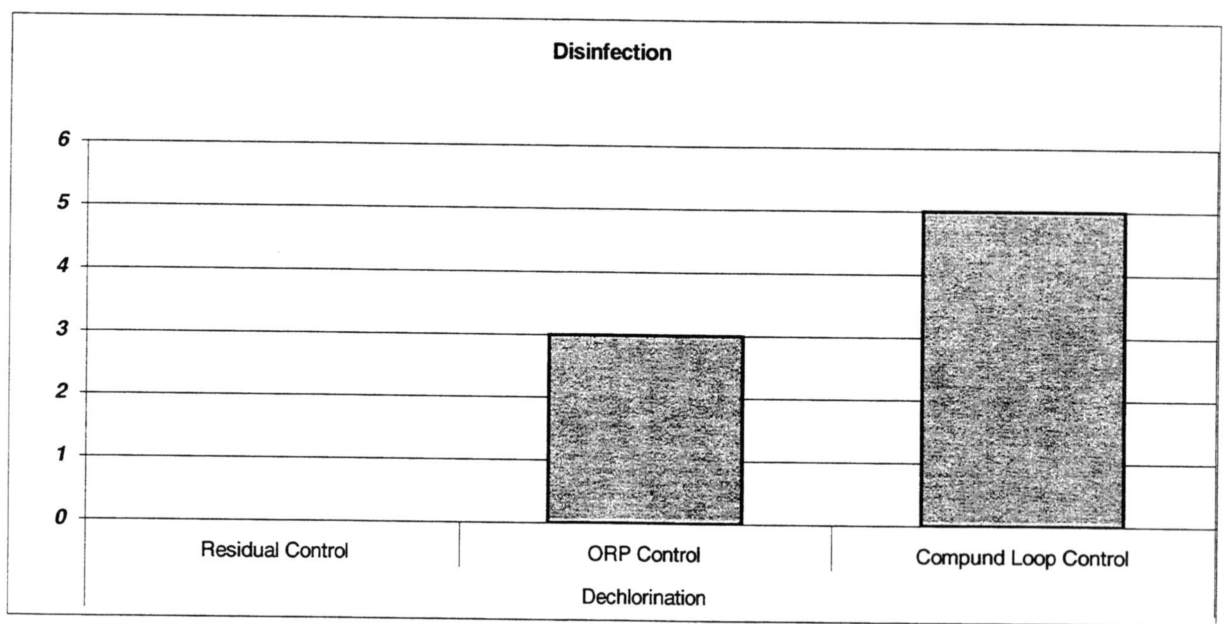

Figure 3-10. Types of Dechlorination Control

UV disinfection systems are typically packaged systems, provided by the lamp supplier. Different manufactures present different approaches to UV control, and provide several options both to lengthen the life of the lamps and reduce energy costs. Four facilities surveyed relied on UV disinfection. Of the four, three used at least two of the different strategies to control UV disinfection (Figure 3-11).

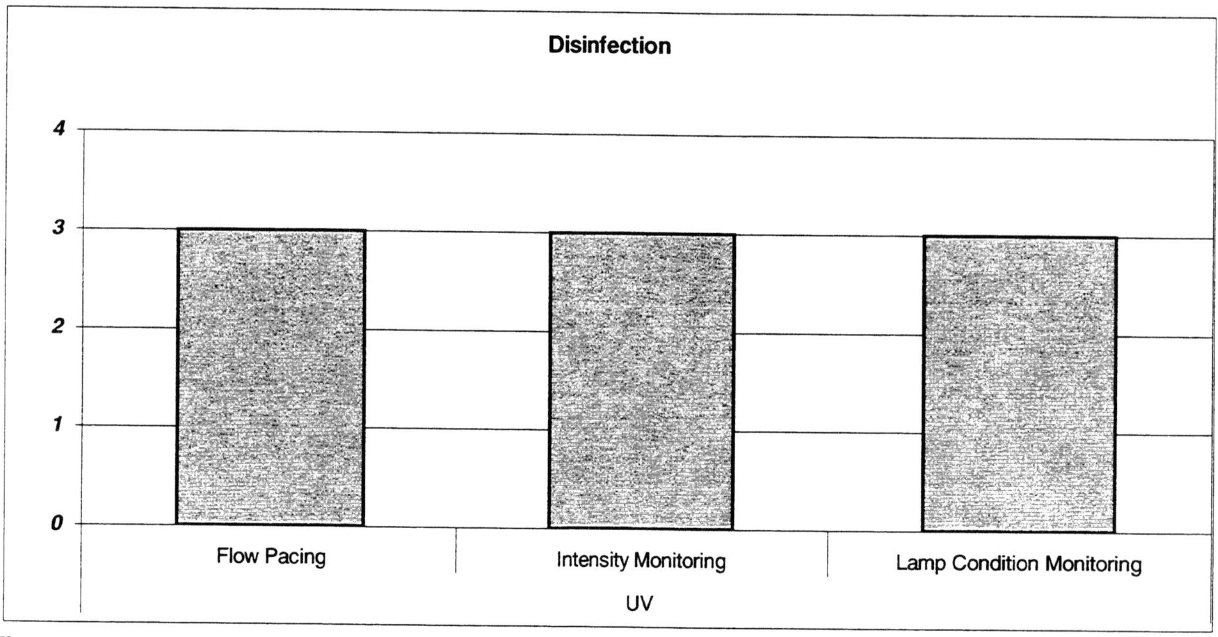

Figure 3-11. Types of UV Disinfection Control

3.3.5 Solids Processing

Unit processes typically associated with solids processing include stabilization (such as anaerobic digestion, aerobic digestion, lime stabilization), thickening, and dewatering. A recent WERF research project (Gillette et al., 2002) suggests that these processes are largely operated with "semi-automatic" controls because of the inherent variations in the feed and the lack of reliable instrumentation. This study concluded that automation is just reaching the solids processing section of the wastewater treatment plant.

The sensor and control system survey generally confirms the conclusion drawn by Gillette et al. (2002). The instrumentation and control systems survey showed that respondents used a relatively low level of instrumentation for monitoring and control of solids processes (see Chapter 2). Physical sensors for parameters such as level, pressure, temperature, flow (liquid, solids or gas), motor speed, vibration (accelerometers), or valve position made up the majority of the instrumentation reportedly in use by the survey respondents. A few facilities reported using more complex instrumentation such as sludge density meters and chemical sensors for pH, conductivity, oxygen, and hydrogen sulphide in their solids processing train.

The WEF publication *Automated Process Control Strategies* (1997) reviewed a number of control strategies applicable to stabilization processes and to thickening and dewatering processes. Stabilization processes addressed in this guidance manual included autothermal thermophilic aerobic digestion (ATAD), a relatively new stabilization process in North America, and anaerobic digestion. Thickening and dewatering processes considered included gravity thickeners, centrifuges, and belt presses.

WEF (1997) noted that ATAD systems are not heavily instrumented and, historically, have been manually controlled. The more recent ATAD designs may include temperature sensors, suspended solids sensors, flow meters, and pH sensors and may incorporate simple automation of the fill-and-draw feed sequence. Temperature control is sometimes accomplished through control of the feed solids concentration or the feed cycle.

Anaerobic digestion of biosolids has been practiced for many years, but, according to WEF (1997), is still largely a manually controlled process. Generally, only temperature, flow, solids concentration, and mixing can be automatically controlled and/or monitored. Flow and mixing control are generally achieved through automation of valves that direct sludge feed to selected digestion tanks and to different locations in the tanks. Temperature control involves the control of heat input to the sludge in heat exchangers, as is common in many industrial processes.

The field surveys identified very few cases where automation had been applied to the control of sludge stabilization processes. Generally, where control was practiced, it was aimed at equalizing the feed rate to multiple primary digesters through manipulation of feed valves.

According to WEF (1997), automated process control strategies associated with gravity thickener operation are very similar to those applied to primary and secondary clarifier operation, that is either sludge blanket level control or solids concentration control

Centrifuges can be used for either sludge thickening or dewatering, and similar control strategies are used for both applications. Control strategies generally applied, according to WEF (1997), have included feed flow control, polymer flow control, and machine control.

Feed flow control is simple but can be effective in improving centrifuge operation. It involves only flow metering instrumentation and a flow control element. Machine control is normally an inherent part of the centrifuge design and varies among centrifuge manufacturers.

Simple polymer flow control involves proportioning the flow of polymer to the flow of sludge to the dewatering or thickening unit based on a pre-determined dosage rate (grams of polymer per liter of sludge). This control strategy is applicable to any dewatering or thickening technology that uses polymers, including gravity thickeners, centrifuges, belt presses, or filter presses. More sophisticated polymer control strategies will proportion polymer dosage to sludge mass flow rate (kg of polymer per tonne of dry solids) rather than volumetric flow rate. This strategy requires a dependable continuous measure of sludge concentration as well as flow.

More sophisticated polymer control strategies vary the polymer dosage in response to a measure of the sludge conditioning or polymer requirements. Various methods of measuring polymer dosage requirements, including zeta potential or streaming current detectors, have been tried with limited success, according to Gillette et al., 2002. Similarly, measurement of the solids or turbidity content of the dewatering centrate or filtrate as a means to control polymer dosage has had limited success (WEF, 1997). WEF (1997) reports on the successful commercialization of a proprietary polymer control system for belt presses that depends on measuring the rheological characteristics of the flocculated sludge. In all cases, successful control of the polymer feed rate depends on accurate and dependable measurement of the polymer requirement.

The field surveys identified two particular examples of gravity thickening control and sludge dewatering control on plate and frame filters that are described in more detail in Chapter 4.

Chapter 4.0

SUCCESSFUL AND PROBLEMATIC PRACTICES

4.1 Introduction and Exclusions

This chapter includes a large body of information on both successful and problematic monitoring and control practices, which will enable users in other facilities to replicate the documented successes for comparable treatment processes. Some of the most useful results of the project were the individual practices noted in the field surveys. In addition, other examples were included from recent literature, facilities known to the project team, and from the WEFTEC 2001 workshop conducted by the project team.

Almost as important as what is included in this chapter is what has not been included. The report is not intended as an overall design guide. It does not include practices that are commonly used and documented in other manuals such as *Instrumentation in Wastewater Treatment Facilities* (1993) and *Automated Process Control Strategies* (1997). There has also been an effort not to overlap with results from other recent WERF projects such as the efforts to evaluate ORP versus chlorine residual control and the control of solids dewatering.

4.2 Process and Instrument Diagrams (P&IDs)

A process and instrument diagram (P&ID) accompanies most of the successful practices documented in this chapter. The P&ID is recognized as one of the most important design documents to convey automation and control concepts and was acknowledged as such in *Automated Process Control Strategies* (1997). The Introduction of that manual defines P&ID and suggests why it is so critical to the design process.

> An excellent tool and starting point for any design is the process and instrumentation diagram. A process and instrumentation diagram is a schematic diagram showing major components of a process or processes, interconnections among them, instruments, measuring points, and control elements. If the

designers can reach a consensus on these major design components, much of the rest of the design is straightforward and simplified.

The Instrument Society of America (ISA) has developed standards for format and content of the P&ID. Specifically ISA standard S5.1, *Instrumentation Symbols and Identification,* presents the standard format for illustrating instrumentation on a P&ID. ISA standard S5.3, *Graphic Symbols for Distributed Control/Shared Display Instrumentation, Logic and Computer Systems,* presents the standards for representing control logic. ISA standard S5.5, *Graphic Symbols for Process Displays,* presents additional graphic symbols intended for use on operator consoles and often used on P&IDs as well.

A licensed control systems engineer should stamp P&ID drawings.

4.3 How To Read a Process and Instrument Diagram

The following is a brief introduction of how to read a P&ID. More complete descriptions can be found in WEF manuals *Process Instrumentation & Control Systems* (1984), *Instrumentation in Wastewater Treatment Facilities* (1993), and *Automated Process Control Strategies* (1997), and the ISA standards. Those not needing the refresher can skip to Section 4.4.

4.3.1 Lower Portion of the P&ID

The bottom portion of the P&ID is a schematic of the process. It shows all the processes, pipes, basins, tanks, vessels, and major pieces of equipment such as pumps, blowers, and valves. All are represented by symbols that are representative and easy to identify. Instruments are shown as "balloons" or "bubbles" with a functional identification and a loop number.

The instrument balloon functional identification consists of two to four letters and is a standardized method of identifying the type of measurement and additional information. The first letter designates the type of measurement. Common first letters include F for flow, P for pressure, "T" for temperature, "L" for level, and "A" for analysis (such as pH, chlorine residual, dissolved oxygen, or ammonia). The following one to three letters can modify the meaning of the first letter and provide additional function information. Table 4-1 lists some common ISA functional identification combinations. Table 4-2 lists the complete ISA standard functional identification letters and their meanings.

Table 4-1. Common Functional Identification Combinations

ISA Identification	Meaning	Example/Explanation
AIT	Analysis indicating transmitter	Dissolved oxygen meter – the instrument balloon subscript identifies the specific analysis
AC	Analyzer controller	Control loop
FE	Flow primary element	Flow meter element
FIT	Flow indicating transmitter	Entire flow meter
HMS	Hand momentary switch	Push button
IT	Current transmitter	Measures electrical current (amps)
LAH	Level alarm high	Tank overflow
LAL	Level alarm low	Pump cutoff level
LC	Level controller	Hardware- or software-based controller
LS	Level switch	Level point measurement
PDIT	Differential pressure indicating transmitter	Used on orifice plates
PI	Pressure indicator	Pressure gauge
TT	Temperature transmitter	Temperature transmitter without a local indicator

Table 4-2. Instrument/Function Identification Letters

	First Letter	Modifier	Input Function	Output Function
A	Analysis		Alarm	
B	Burner, Combustion			Close, Stop, Decrease
C	Communication			Control
D		Differential	Direction	Open, Start, Increase
E	Voltage		Sensor (Primary Element)	
F	Flow	Ratio (fraction)	Fail	
G			Glass, Viewing Device	
H	Hand		High (Opened)	
I	Current (Elec.)		Indicate	
J	Power	Scan		
K	Time	Time Rate of Change		Control Station
L	Level		Low (Closed)	Light
M	Motor, Pump	Momentary	Middle (Intermediate)	
N	Intrusion		On, Operating	
O			Orifice, Overload	
P	Pressure, Vacuum		Point (Test)	
Q	Quantity	Integrate, Totalize		
R	Radiation		Record	
S	Speed, Frequency	Safety		Switch
T	Temperature			Transmit
U				
V	Vibration, Mech. Analysis			
W	Weight, Force		Well	
X	Unclassified	X Axis	Unclassified	Unclassified
Y	Event, State, Presence	Y Axis	Relay, Compute, Convert	
Z	Position, Dimension	Z Axis		

The loop number consists of one or more numbers and uniquely identifies the instrument balloon or function. Many designers further imbed geographic or process information into the

loop number by designating the first one or two numbers as a process area and the following numbers as sequential identifiers.

Instrument signals are shown with dashed lines that generally lead towards the PLC or other controller. Analog inputs (for examples, DO or flow) and outputs (for example, valve position) are shown with a solid arrowhead while discrete points (switches) are shown with hollow arrowheads.

4.3.2 Upper Portion of the P&ID

The upper portion of the P&ID shows the controller and the logic that resides within. For the purpose of illustration in this report, all controllers are shown as programmable logic controllers (PLCs). Relations between signals are generally shown as arrows going from the source to a destination. A diamond with an "I" in it is a generic reference to an interlock. The interlock is usually described in more detail in another document such as an electrical control schematic or process control narrative.

A diamond with an identification ending in "C" (for example, AIC) signifies a control loop. One input to the control loop may be designated as "SP" meaning the setpoint. Another input into the control loop may be designated as "PV" meaning process variable. The control loop will attempt to manipulate its output to maintain the process variable at the value of the setpoint. Other diamonds or squares with mathematical symbols within them (Σ for summation, Δ for difference, S/H for sample and hold, etc.) signify that specific mathematical function.

4.3.3. Simplifications Made for Report P&IDs

Several simplifications have been made to the P&IDs presented in this report. Generally no local control panels, area control panels, or motor control centers are shown. While these components are critical for true design drawings, they generally complicate the drawing and are not strictly needed to demonstrate the control concepts.

4.3.4 Other Design Documents That Supplement the P&ID

P&IDs, while one of the most important documents, are also complementary to other forms of process control logic documentation and may be supplemented by flow charts, logic diagrams, process control narratives, elementary wiring diagrams, loop drawings, and other written documents and drawings. These other documents will generally use the names and references shown on the P&IDs.

4.4 Unit Process Based Practices

During the course of conducting this research, the authors found a large number of control strategies not commonly documented in current literature that had been successfully applied. The following sections document these control strategies on a unit process basis. Critical components, as reported by the owner, are documented. When costs and savings were known, these values are also reported. However, the majority of utilities documented neither costs nor savings. Commonly used control strategies available from other readily available sources are not reported.

4.5 Influent Lift Stations, Preliminary Treatment, and Primary Treatment

4.5.1 Deep Tunnel Pumping System – Boston, MA

Hydraulic configuration and process conditions can drastically complicate a pumping system. At Deer Island WWTP in Boston, Massachusetts, the raw wastewater flows from the northern metropolitan area to the treatment plant via two long, deep tunnels. One tunnel is fed from a single shaft and headworks. The other tunnel is fed from two different shafts and headworks. (For the purpose of this report, the single headworks strategy is discussed. The two headworks feeding a single tunnel is a more complex version of the strategy and is discussed in the reference article.)

The shafts are approximately 260 feet deep, and the tunnels are ten feet in diameter. A pump station at the wastewater treatment plant draws raw wastewater from the shafts and pumps it into the plant. The pump station houses 10-3500 hp (165 mgd) pumps. The tunnels are constructed with a nominal downward grade of 1.5 feet per mile, but the overall operation depends on matching the pumped flow out of each tunnel with the flow into the tunnel, at the headworks. The pump station at the wastewater treatment plant serves both deep tunnels, so pumps must be configured to pump from either shaft.

Process instability results from several hydraulic features. First, the shafts and tunnel are similar to a U-tube; any differences between flow in and flow out create the potential for developing oscillations. The U-tube has a period resembling a pendulum (Equation 4-1).

$$\tau = 2\pi\sqrt{\frac{L}{2g}} \qquad [4\text{-}1]$$

Where τ is the period, L is the distance between unbounded surfaces, and g is gravitational acceleration (ft/s^2). Based on the hydraulic characteristics, one tunnel has a 114-second period, the other a 153-second period. Head loss through the tunnel causes some damping, and the control system applies additional damping such that the period of the observed cycles is between 3 and 5 minutes.

To further complicate the control challenge, the fully rated pumped flow of 750 mgd can empty the shaft at Deer Island in only 4.5 seconds.

To address this process challenge the control strategy is to maintain a constant level at the headworks facility. Due to the process dead time, and the resonant process, with the potential for developing oscillations, the controls are far more complex than a simple level controller to maintain the headworks level at a fixed setpoint.

The control strategy developed to solve this complex process challenge has evolved over the years. The pump station and tunnels were initially constructed in the 1960s, and a control strategy was developed to address the controls at that time. With the recent construction of the new Deer Island Treatment Plant and retrofit of the pump station, the control strategy was revisited and revised.

Flow through is controlled by calculating a head difference equal to hydraulic losses in the tunnel and pumping to maintain that level. The hydraulic loss is calculated as a function of the square of the process flow, proportional to losses due to head loss (Equation 4-2).

$$h_2^* = h_1^* - k_1 F_1^2 \qquad [4\text{-}2]$$

Where h_2^* is the setpoint level needed at the pump station in order to keep the level at the headworks constant, h_1^* is the level at the headworks, F_1 is the flow through the tunnel, and k_1 is the head loss coefficient. The headloss coefficient is presumably constant and can be field adjusted to set the desired level for the headworks. For example, the head loss coefficient for the tunnel in discussion was calculated at 30 feet for maximum pump station flow and held essentially constant for lower flows.

The flow signal biases the output of the level controller, preventing the need for integral action in the controller.

$$F_2^* = F_1 + G_2(h_2 - h_2^*) \qquad [4\text{-}3]$$

Where F_2^* is the pumped flow setpoint, h_2 is the measured shaft level, and G_2 is the gain of the level controller. The gain setting has an optimum value, which produces the maximum damping for the loop. Lead-lag compensation on the headworks flow signal is required to accelerate the signal to get the entire control system moving.

The P&ID illustrating the control system is shown in Figure 4-1.

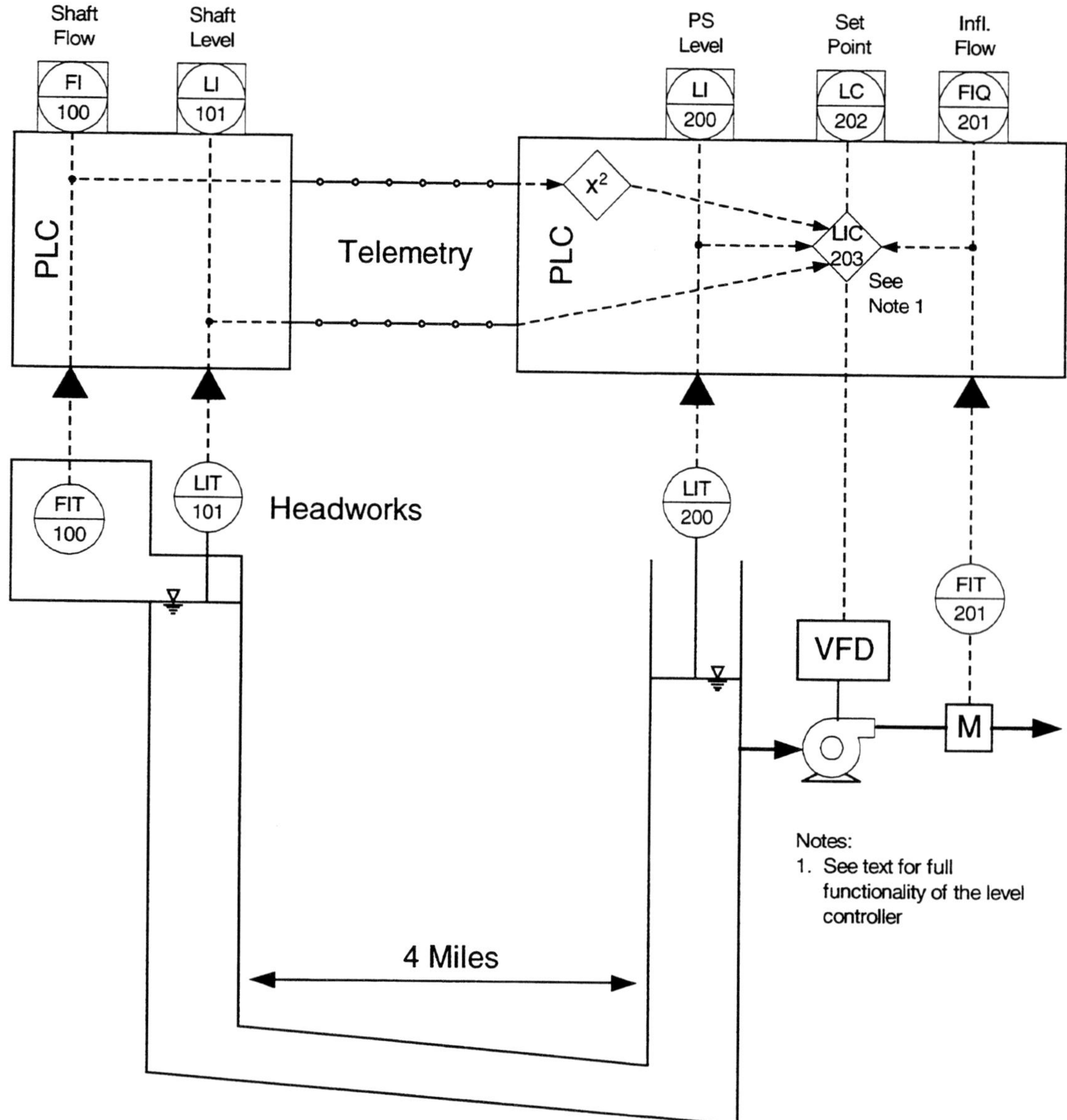

Figure 4-1. P&ID for the Deer Island Pump Station Controller

Several real world startup challenges complicated the startup process. The level measurements at the headworks and at the pump station, miles apart, did not initially have the same zero value. Calibration of the instruments took some effort because the instruments were located at such a great distance from each other, and setting them to the same zero required stopping all flow through the pump station until the flow damped and the values could be set the same.

During periods of low flow, solids settled in the tunnel, increasing headloss. During a storm surge, the solids became resuspended and changed the headloss characteristic again. The headloss coefficient needed some type of on-line calibration to adjust for the change in solids deposition. The method used to recalibrate the strategy relied on operator intervention. When the tunnel level reflected an unacceptable level of offset from the predicted level during high, steady-state flow, the operator manually adjusted the coefficient to correct for the offset.

Finally, the telemetry used to transmit headworks flow and level signals were configured to update only when the process changed by more than 2% and only every 5 minutes. This introduced unacceptable dead time into the controls, making the process less stable. The only way to correct for the dead time was to replace the telemetry.

The controls automatically set the pump speed of the 3500 hp pumps, but the starting and stopping required operator action. When the control strategy called for a pump to start, all the other pumps would be running at full speed. The operator would start the next pump, which would ramp up to full speed prior to lifting its check valve. The resulting surge in flow caused significant process instability. To correct for it, the automatic controls would ramp all running pumps down to the correct speed for the new number of pumps on line prior to starting the next pump. This would minimize the surge and the resulting process upset.

All the steps required to make this complex control strategy operational took a significant level of effort from a highly knowledgeable group of experts. The strategy was developed by F.G. Shinskey (1985), modeled by him, and fully tested theoretically prior to implementation. Shinskey was also involved in testing and tuning the control strategy for optimum operation. His involvement and knowledge was critical to successful implementation.

4.5.2 Influent Pumping Flow Optimization/Prediction –Tillsonburg, Ontario

Hydraulic disturbances or flow fluctuations have been shown to have substantial and detrimental effects on the removal efficiencies of both primary and secondary settlers. These disturbances can result from variations of the influent flow, operation of the influent and recycle pumps, and external factors such as wind.

Based on modeling with a turbulent diffusion model, Collins and Crosby (1980) concluded that flow disturbances have large effects that persist longer than the actual disturbance. In a full-scale secondary clarifier, Pflanz (1969) observed effluent solids varying both with hydraulic loading and solids loading. In a full-scale primary clarifier, Olsson (1978) observed that a 25% increase in flow rate for 0.5 hour resulted in almost doubling the effluent solids. In a pilot-scale activated sludge system, Chapman (1983) observed that increases in flow were accompanied by immediate increases in effluent solids that decayed slowly. Hill (1985) observed similar results in a full-scale activated sludge system.

In general, variations in influent flow rate result in higher average primary solids and loading to the biological processes and higher average effluent suspended solids concentrations than steady flows.

The minimum effluent suspended solids concentration for a given volume of wastewater is attained with a constant flow rate. A constant flow rate, however, requires a substantial equalization volume that is not usually available at municipal treatment plants. What is available is a smaller volume of influent wet well volume that can be used to minimize the flow fluctuations. For any given influent pattern and a finite equalization volume, there exists an optimal outflow that will minimize the flow fluctuations and, thereby, maximize the treatment efficiency. This concept is shown in Figure 4-2 and was implemented by Newbigging (1987).

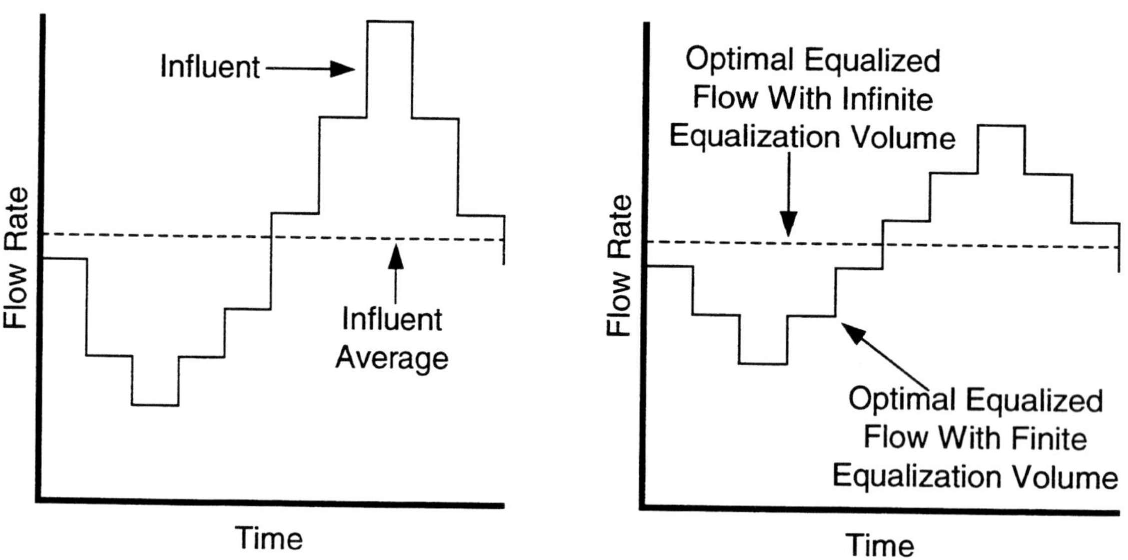

Figure 4-2. Flow Equalization with Infinite and Finite Volumes

Figure 4-3 is an overall schematic of the lift station controller implemented by Newbigging (1987). It consists of several components including a flow forecaster, a dynamic model of the treatment plant, and a dynamic optimization routine. The dynamic optimization process is in overall control and attempts to make optimal control decisions by trying multiple outflow patterns and evaluating the response of the simulator until preset conditions are met. Some of the preset conditions are that the wet well can be neither drained nor overflowed and that the pump capacity can not be exceeded. The simulator uses inputs from the flow forecaster and the outflow from the optimization routine to predict effluent solids and lift station levels. The influent flow forecaster uses an autoregressive time series model and actual flow measurements to predict the influent flow some several hours into the future.

Newbigging (1987) reported that the results of the flow forecast algorithm lift station controller produced better results than either an on/off controller or a level controller. Only the flow forecast controller produced better results given a larger wet well volume.

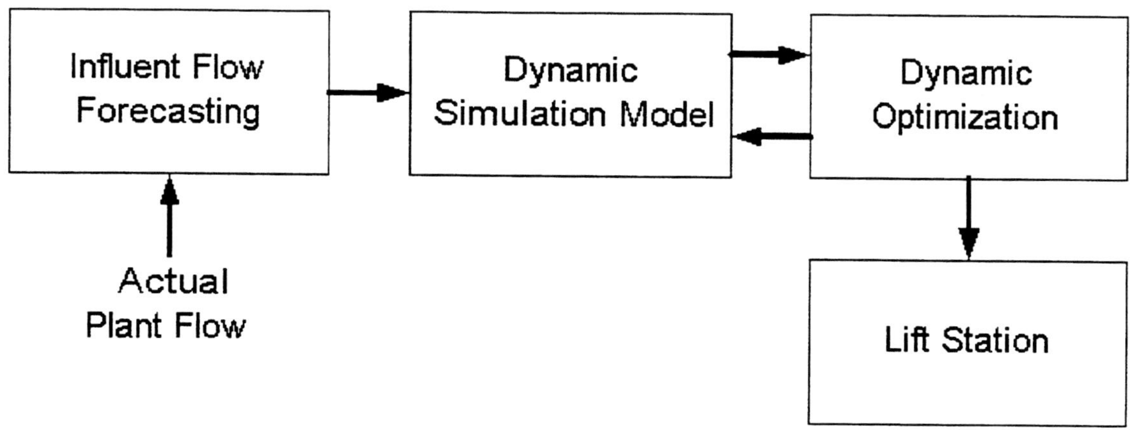

Figure 4-3. Structure of the Optimal Lift Station Controller for Smoothing Flow

4.5.3 Screening Control – Boston, MA

The bar screens at Boston's Deer Island WWTP utilized a unique automatic rake operation based on differential level control principles but without a differential level measurement.

Bar screen rake operation based on timed operation, where the operator manually entered the rake operation frequency, resulted in frequent mechanical breakdowns and unreliable operation. The operators would place the rake on continuous operation in preparation for the occasional flooding condition. The frequent operation caused significant wear and tear on the screen, resulting in frequent breakdowns and high maintenance costs.

The screens were scheduled for replacement, so any significant changes to the instrumentation or mechanics were not practical. Instead, a controls-based solution was implemented. Each time a rake operation finished, the control strategy would sample-and-hold the upstream, clean-screen, channel elevation and start an elapsed-time-since-last-rake-operation timer. The control strategy continuously compared the clean-screen channel level with the current channel measurement, and the amount of time until then next rake operation was calculated continuously. When the elapsed time exceeded the calculated time to next operation, a rake operation was initiated, and the sequence started again.

The time until next rake operation was calculated based on the starting clean-screen level. For example, when the measured level was the same as the clean screen level, the time between next rake operation was 3 hours, when the level was 10 inches higher than the clean screen level, the time between rake operation was reduced to 30 minutes.

The Table 4-3 summarizes the initial settings.

Table 4-3. Initial Rake Frequency Settings for Differential Level Screen Operation

Difference in Level (Current Level – Last Clean-Screen Level) [inches]	Rake Frequency [minutes]
0	180
5	60
10	30
20	15
50	5
100	5

The time until next rake operation was continuously calculated as a function of the rising channel level. If the level didn't change at all, then the rake would not operate until the total time had elapsed. When the elevation started to increase, the time until next rake decreased proportionally. This type of strategy did cause the rake to operate more frequently during high flows (high levels) and the leading edge of a storm, when the level rose due to increased flows. This was not viewed as an issue, because the leading edge of a storm flow typically requires frequent screening.

The strategy included a continuous operation mode. When the clean-screen channel elevation exceeded a high level setpoint of 35 inches, the strategy switched the rake to operate every 15 minutes for as long as the level stayed high.

The P&ID illustrating the control system for the Deer Island bar screen is shown in Figure 4-4.

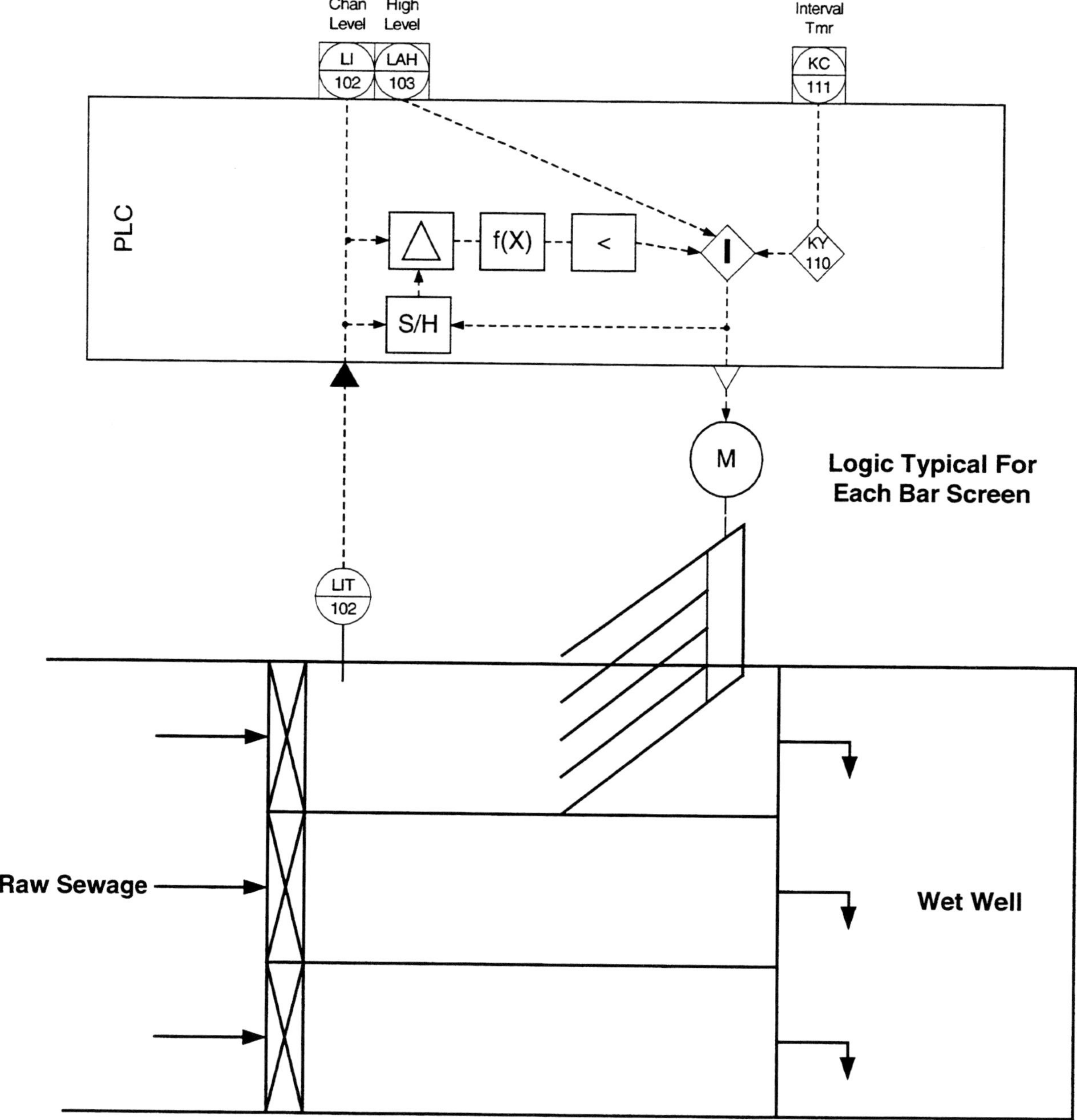

Figure 4-4. P&ID for Deer Island Bar Screen Controller

4.5.4 Automatic Grit Channel Velocity Control – Milwaukee MSD South Shore WWTP

Grit removal is accomplished in four grit channels. To accommodate highly variable plant flows under wet weather conditions and the need to have grit channels out-of-service for cleaning, an automatic velocity control system was implemented to maintain an optimum range of velocity for grit removal. Key elements of the system include:

- an ultrasonic level sensor in each grit channel to measure level;

- a magnetic flow meter on the discharge from each grit channel; and,
- a butterfly valve to restrict flow from each grit channel.

The P&ID illustrating the control system is presented in Figure 4-5. The level in each grit channel and the grit channel flow is measured continuously. The flow velocity (ft/sec) is calculated by the control system based on the flow, level, and the channel dimensions. The control system automatically adjusts the downstream butterfly valve to control the flow and level and maintain the preset velocity in the grit channel at between 0.8 and 1.25 ft/sec.

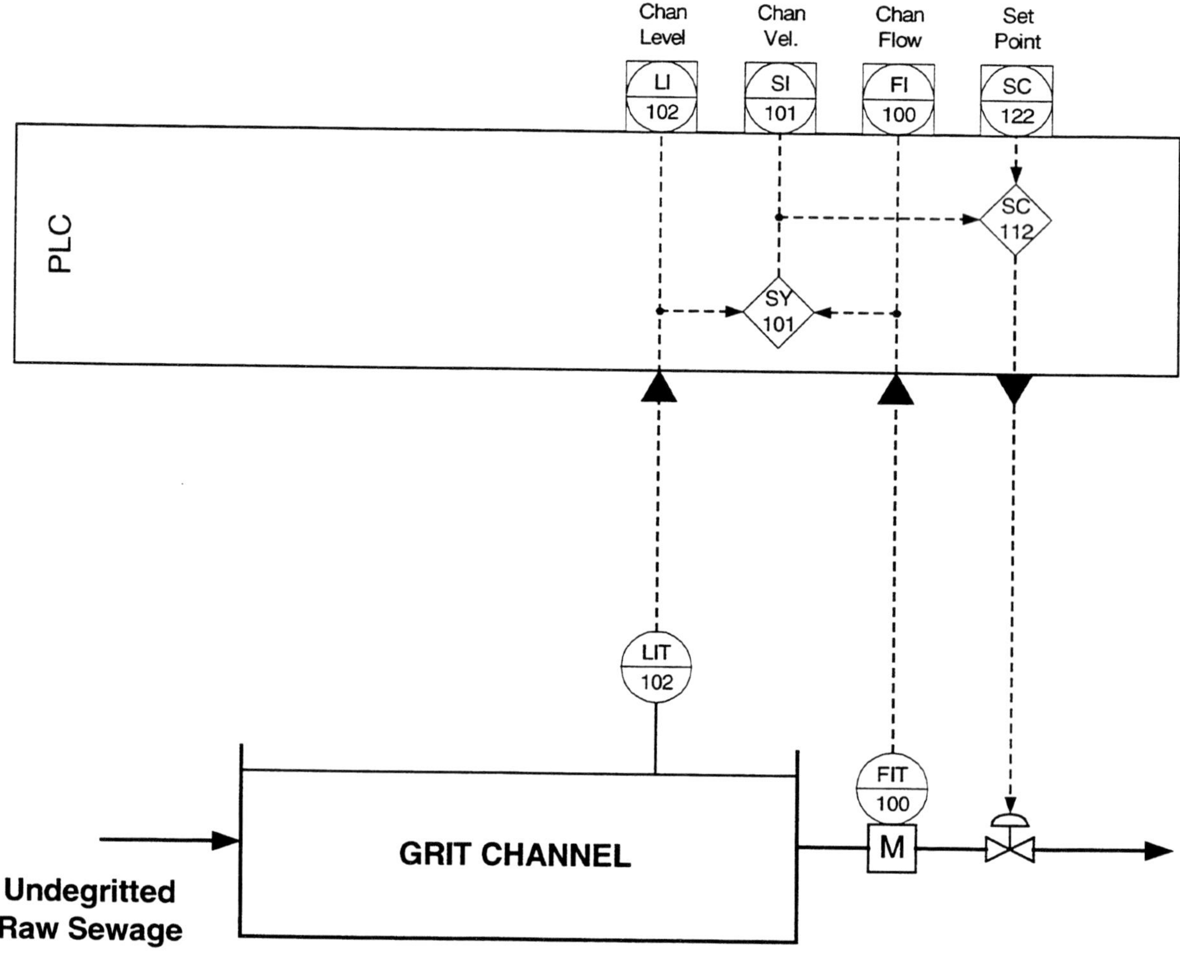

Figure 4-5. P&ID for Grit Channel Velocity Control

4.6 Secondary Treatment

4.6.1 Aeration (Dissolved Oxygen) Control

Automatic aeration (or dissolved oxygen) control has been utilized in wastewater treatment systems with varying degrees of success for at least the past 30 years. The U.S. EPA (Flanagan, 1977) published a design manual on the topic 25 years ago. Aeration control has been demonstrated to provide process improvements in most instances by providing the right amount of aeration at the right place and right time. Aeration control has been consistently demonstrated to provide 15% to 30% savings in electrical costs for most municipal facilities compared to manual control.

The WEF manual *Automated Process Control Strategies* (1997) presented several aeration control strategies of varying complexities. No new control strategies were observed in this research. Figure 4-6 shows a typical aeration control strategy.

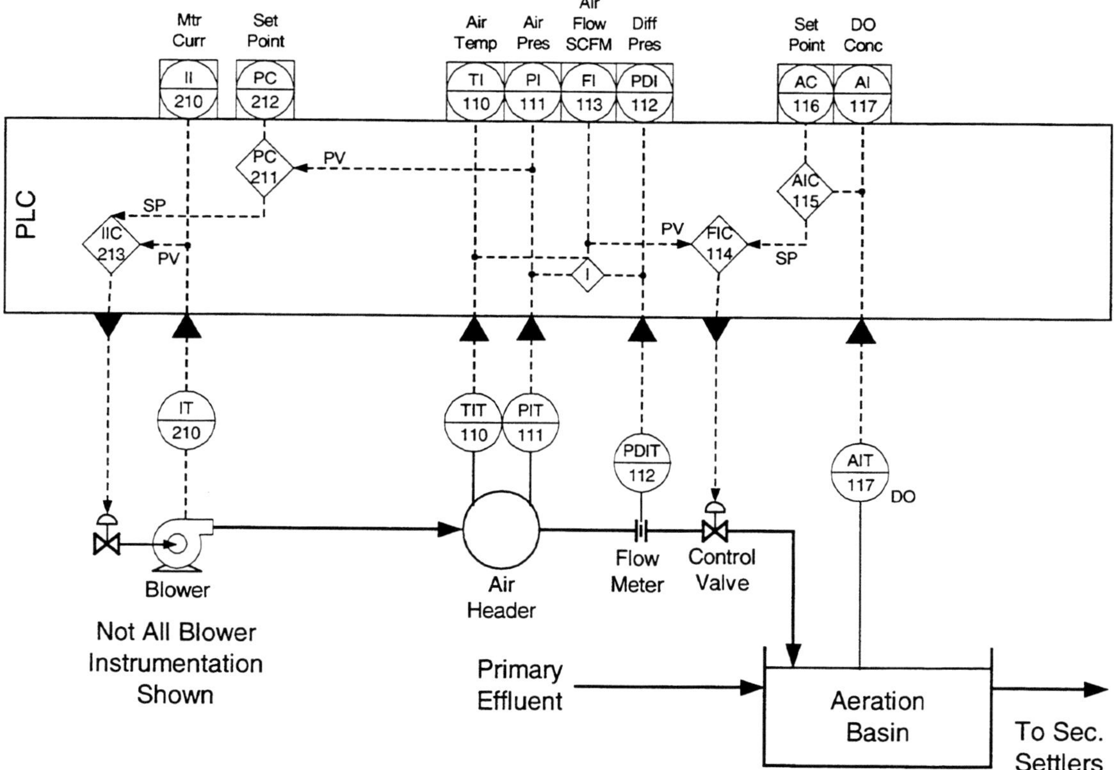

Figure 4-6. P&ID for Aeration Control

In order to meet the different time-varying air demands of the individual aeration basins, as well as for the system as a whole, aeration control is divided into air distribution control and total air production control. The air distribution controller permits control of each aeration basin at the desired DO setpoint. The purpose of air production control is to provide the correct amount of total air such that each aeration basin can meet its DO setpoint.

The control of air distribution is shown on the right side of Figure 4-6. Two cascaded control loops are required for each aeration basin. The outside loop (AIC-115) controls DO. Its setpoint (AC-116) is typically an operator-entered value, although some strategies allow a time-varying setpoint that is the result of a calculation. Its process variable is the actual DO concentration. Its output is a desired airflow rate that is used as the setpoint to the airflow loop (FIC-114). The airflow control loop uses the calculated value of mass flow rate (FI-113) as its process variable and outputs the airflow control valve position.

To see how this strategy works, assume that the organic loading on the aeration basin increases, which will, in turn, increase the biological oxygen uptake and cause the DO concentration to decrease. As DO goes down, the DO controller (AIC-115) will increase its output to the airflow controller. The airflow controller will open the flow control valve and increase the airflow rate. Increased airflow will increase the DO concentration and the system will tend towards meeting the DO setpoint again.

Timing of how often the two-control loops run is essential to their correct operation. Timing should be correlated to how fast the physical system reacts. For instance, when a step change in airflow rate is made, it may take 30 minutes or more for the full effect to be reflected in the DO concentration. DO typically has a time constant (63% of ultimate change) of about 15 minutes. Control loops usually run 3 to 10 times per time constant. Therefore, the DO control loop should run every 3 to 5 minutes. The airflow, however, responds much faster to changes in pressure and valve position and has a time constant of seconds. The airflow loops typically run every 15 to 30 seconds.

Air distribution control systems must also implement other constraints such as minimum and maximum airflow rates to each basin. A minimum airflow rate is needed to provide at least minimum mixing and prevent solids from settling in the aeration basin. Systems with fine bubble diffusers usually have both minimum and maximum airflow rates to protect the diffusers. Some manufacturers recommend values of 0.24 to 1.42 l/s (0.5 to 3.0 scfm) per diffuser.

Modern digital controllers permit assigning both minimum and maximum outputs from each control loop. These limits prevent damage to the process or equipment in the event of sensor failure. Thus, even in the event of complete sensor failure, the control loop will only travel to the extent of the output limits.

Air distribution control as described above works so long as the system header pressure is sufficiently high that no control valves are completely open and not meeting their airflow setpoint. If the pressure is too low, the DO loop will call for more airflow, but the airflow loop will not be able to open the valve any more to provide it. In this case, the blowers must provide more air to raise the system header pressure by opening their inlet guide vanes or starting another blower. One potential method for coupling the air distribution and air production control systems is discussed in the next section.

Air production control consists of modulating the blower output to produce the appropriate amount of air at the minimum cost. One method for accomplishing this objective is shown on the left side of Figure 4-6. This "traditional" means of regulating total airflow rate is to control at a constant system header pressure. If the measured header pressure is too low, the blower output is increased until the pressure is again raised to the value of the setpoint. Likewise, if the pressure is too high, the blower output is lowered. In this manner, a wide range of airflow rates can be obtained at a constant pressure.

Figure 4-6 shows a centrifugal blower being controlled by two cascaded control loops. The outer loop (PC-211) uses the system header pressure as its process variable and outputs a setpoint to the inner current loop. The inner loop (IIC-213) measures blower current and modulates the inlet guide vanes. The pressure loop typically runs at an interval of 4 to 5 seconds while the inner loop executes about once per second. Current is used for the inner loop because it is an indicator of the amount of work being produced and a good indicator of the airflow rate. Current is also useful in preventing centrifugal blower surge.

Positive displacement blowers can use a similar control strategy with a variable frequency drive (VFD) used instead of an inlet guide vane to provide a variable airflow rate.

It is almost always desirable to produce the required amount of air at the minimum cost. This objective can usually be met by minimizing the system header pressure. System pressure is minimized when at least one control valve in the distribution system is fully open (or nearly so). Therefore, many control systems utilize an algorithm that attempts to maintain at least one distribution control valve nearly full open by periodically adjusting the system header pressure setpoint. This algorithm is known as the "most open valve strategy."

Despite the demonstrated process improvements, the substantial financial rewards, and the wealth of design and operating information, aeration control still is not universally applied to activated sludge systems. In fact, it is still the exception. One reason for the lack of acceptance is that a properly functioning aeration systems requires almost every component in the system to be properly designed and operating correctly. In practice, this is seldom the case. Some of the commonly observed problems are:

- ♦ Air lines are oversized for the actual airflow rates and result in low air velocities. The low velocities lead to airflow meters operating in the bottom 10% to 20% of their range with poor accuracy.
- ♦ Airflow control valves are often the same size as the air line and oversized. This leads to the valves operating less than 20% open where there is poor control authority.
- ♦ Aeration blowers are often oversized. If a centrifugal blower is throttled as much as possible or a positive displacement blower is operating at minimum speed and still exceeding the air requirements of the plant, there can be no effective control and no power savings.
- ♦ Aeration blowers often have limited turn-down capacity. If each blower cannot be turned down to less than 50% of capacity, a control gap exists between the output of one blower operating at maximum capacity and two blowers each operating at minimum capacity. Such control gaps lead to poor control and potentially to excessive cycling of the blowers.
- ♦ Dissolved oxygen probes are often located at the end of the last pass of the aeration basins. Due to detention time in the aeration basins, this location is often 6 to 12 hours out of phase with the incoming load and, consequently, a poor location upon which to base aeration control.
- ♦ Dissolved oxygen probes, airflow meters, and valve actuators are often mounted in difficult or dangerous locations with little access for maintenance. Instruments not

designed with maintenance in mind are often not maintained and seldom perform reliably well. Instruments such as DO probes need periodic cleaning and should ideally be mounted so that no tools are needed for routine access.

4.6.2 Respirometry for Aeration Control at an Industrial Treatment Plant – Blue Ridge Paper, Canton, NC

The Blue Ridge Paper treatment plant uses a conventional activated sludge process to treat a waste that is 97% industrial and 3% domestic. It has a design capacity of 2.63 cubic meters per second (60 mgd) but currently treats an average of 1.10 cubic meters per second (25 mgd). The plant previously operated near capacity but now has considerable excess capacity due to reductions in waste from in-plant modifications. The plant is lightly instrumented and currently runs almost entirely in manual control. However, from 1993 through 1999, the plant used two respirometers to control aeration. The plant has surface aerators that can be run at high speed or low speed or can be turned off.

Two respirometers were installed in 1993 for $132,000. One measured respiration rate on the primary effluent (influent to the activated sludge process) and the other was used on mixed liquor (MLSS). The general strategy was to determine the load coming into the plant and adjust aeration capacity accordingly to match. The MLSS measurement was made to ensure that the treatment system was not overloaded and the microorganisms were in an endogenous phase at the end of aeration. The plant operators used the respirometers as indicators and performed all control actions manually.

The respirometers functioned well and ran relatively trouble free once properly installed; however, considerable problems were noted during startup. The most significant problem during normal operations was clogging of the sample lines. The sample lines were approximately 200 meters long and were insulated. Any foam in the process resulting in clogging the lines and shutting down the instruments. The operators noted that it was difficult to keep a constant flow rate to the instruments. On good days, the respirometers took only about ½ hour per day to maintain. However, on other days, maintenance might take as much as 3 to 4 hours.

The respirometers were reported to be useful in determining whether the plant was overloaded and needed additional aeration capacity. They were not successful in detecting spills because the foam associated with such spills would clog the sample lines and shut down the instruments. Conductivity monitors in the plant sewers were ultimately used to detect liquor spills from the plant.

When the plant manager retired in 1999, the respirometer project was abandoned after six years of operation. At about this same time the instruments started experiencing failures due to corrosion of their printed circuit boards. The corrosion problem was reported as being the result of poor quality instrument air. The current plant manager believes DO measurements are a better indication of plant overloading.

The staff reported that power savings over a six-month period exceeded the installed cost of the instruments although no hard data is available. The mill users have never published a paper on the system, although many other companies have visited the facility to see the respirometers in operation.

4.6.3 Influent Load Control at an Industrial Treatment Plant – DuPont, Charleston, SC

The treatment plant at DuPont's Cooper River site treats industrial waste generated at the site from fiber production. The treatment process consists of an equalization basin followed by activated sludge. The plant also has a diversion tank that may be used to store influent too strong organically to treat. Flow to the diversion tank is based on a total organic carbon (TOC) concentration as a measure of organic material. If the influent TOC concentration exceeds a preset limit that might upset the treatment process, the influent flow is routed to the diversion tank rather than the equalization tank. Wastewater in the diversion tank is manually bled back into the treatment plant at times of lower organic loading.

The control strategy depends on the reliable operation of the TOC instrument. This particular instrument is based on an ultraviolet-promoted persulfate oxidation principle in which soluble organic matter in the water is oxidized to carbon dioxide and measured with an IR detector in a gas phase. The instrument needs a sample relatively free of solids and requires filtration pretreatment. Dupont uses a simple cartridge filter that must be changed weekly. Changing the filter and adding chemicals only takes a few minutes each week. The control system also sends out an alarm when the TOC value gets too low, since this usually represents a loss of flow due to clogging.

The DuPont plant also used its own unique flow-through cell for pH measurements. The cell held the pH probe and a thermometer for standardization. The flow-through cell could be valved-off, drained, and filled with pH standards for in-place calibration. The pH probe was cleaned and standardized weekly.

4.6.4 Automatic Sludge Wasting Control

4.6.4.1 Importance of Sludge Wasting

Control of the sludge wasting rate from an activated sludge process is considered one of the most important control parameters. Manipulating the wasting rate has the overall objective of influencing the biology of the activated sludge process. Although the biology can be affected by influent conditions and other operational variables such as dissolved oxygen concentration, the wasting rate is generally considered to be the most influential variable. This conclusion is supported both by mathematical analysis of microbial growth and by practical experience in the field over many years.

Although the sludge age concept is used at many facilities, automatic control of sludge wasting based on sludge age is not yet commonly used. Clearly there is a potential for improved benefits from controlling the plant more closely. Benefits attributed to automatic sludge age control in the WEF manual *Automated Process Control Strategies* (1997) include reductions in permit violations, increases in design capacity, control of nitrification, and control of filamentous organisms in the aeration basins.

Automatic sludge age control was evaluated at five different facilities. The five plants used four different types of sludge age control with one plant implementing two different techniques. Each plant had different levels of instrumentation and technical complexity. The next section describes each automatic sludge age control strategy implementation; develops the implicit and explicit assumptions; and presents advantages and limitations of each.

4.6.4.2 Sludge Age Control

The concept of sludge age control is well known and discussed in *Automated Process Control Strategies* (1997) and *Instrumentation in Wastewater Treatment Facilities* (1993) and numerous textbooks. By definition, sludge age is defined as the average time solids are retained in the activated sludge process. A practical definition is the mass of sludge in the aeration basins divided by the amount of sludge wasted intentionally and in the effluent as shown in Figure 4-7 and Equation 4-4.

$$SRT = \frac{(V) \times (MLSS)}{QwXw + QeXe} \quad [1]$$

[4-4]

where: SRT = solids retention time or sludge age (T)

V = volume of aeration basin (L^3)

MLSS = mixed liquor suspended solids concentration (M/L^3)

Qw = waste flow rate (L^3/T)

Xw = solids concentration of waste stream (M/L^3)

Qe = effluent flow rate (L^3/T)

Xe = solids concentration of secondary effluent (M/L^3)

If the secondary clarifiers contain a substantial portion of the solids in the activated sludge process, the numerator of Equation 4-4 should be modified to include the total amount of solids in the system including those in the clarifier.

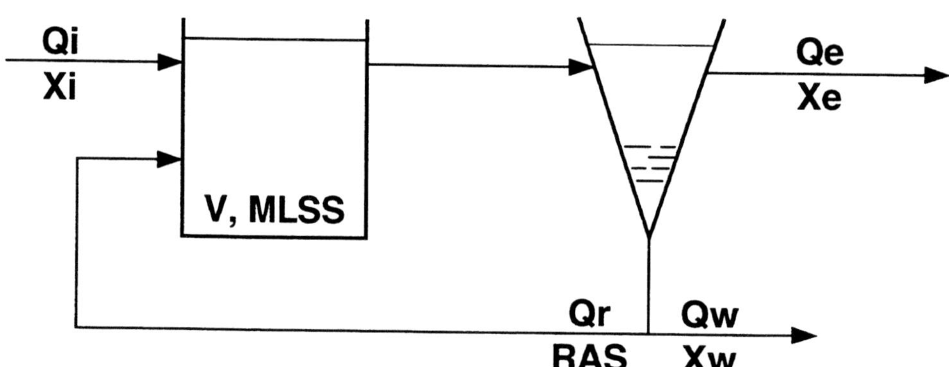

Figure 4-7. The Activated Sludge Process

Equation 4-4 can be rearranged to solved for the waste flow rate as shown in Equation 4-5. Of the two terms, the first term (associated with intentional wasting) is most important. The second term (associated with unintentional wasting in the secondary effluent) is usually small (<5%) for well-operated activated sludge systems and can often be estimated or simply ignored.

A possible exception to this is very high sludge age systems, such as those often used for industrial wastes or extended aeration municipal systems.

$$Qw = \frac{(V) \times (MLSS)}{SRT \times Xw} - \frac{QeXe}{Xw} \quad [2]$$

[4-5]

Fully implementing Equation 4-5 requires three solids concentration measurements (MLSS, RAS, Xe) and at least two flow rate measurements (Qi or Qe, Qw).

4.6.4.3 Five Case Histories

Each automatic sludge age controller discussed below uses some variation of the basic sludge age concept. Each, however, is unique in the requirements for on-line measurements and calculations. Each also makes assumptions that may or may not apply to other facilities.

San Jose, California

In San Jose, the basic sludge wasting equation (Equation 4-5) is almost directly implemented with on-line measurements of flows, MLSS, and RAS. The effluent solids term is assumed to be insignificant and is simply ignored. Two challenges San Jose faced were how to accurately calibrate the suspended solids analyzers and how to account for periodic instrument problems.

San Jose solved its calibration challenge by using a portable solids probe to calibrate each process instrument. The portable instrument was calibrated 1 to 4 times per month with laboratory data. It was then used to adjust the calibration on all of the process solids analyzers in the field that were actually used in the control strategy, ensuring that all analyzers were calibrated to the same standard and minimizing laboratory sampling.

San Jose solved the challenge of periodic inaccurate analyzer data by implementing a statistical method to look at the solids signal and determine when it was seeing “bad” values. Often the instrument problem was really a problem with the sample pumping system. When the procedure determined that the analyzer was sending poor quality data, safeguards were implemented to prevent the solids wasting algorithm from going out of reasonable bounds. An alarm would also sound so that the problem could be corrected.

One interesting result of this strategy was that it was implemented into a commercial instrument that is on the market today.

The advantages of the San Jose strategy include:

- Improved SRT control compared to the manual method previously used. San Jose experienced greatly reduced filamentous foaming after implementing automatic SRT control.
- Reduced laboratory analyses by 80%.
- Reduced variability of WAS to downstream unit processes and improved performance of DAF sludge thickening.
- Data quality algorithm correctly identified and corrected for instrument and sampling errors.

Tillsonburg, Ontario, Canada

Automatic sludge age control was implemented in Tillsonburg, Ontario as part of an automation demonstration project conducted by the Wastewater Technology Centre. The Tillsonburg implementation is unique in that a simulation model was used to calculate the MLSS concentration and eliminate the need for one solids instrument. Another unexpected benefit was that the modeling technique eliminated the need to accurately calibrate the one remaining solids analyzer.

The Tillsonburg automatic sludge age controller was also based on Equation 4-5. However, rather than measuring the MLSS concentration with a solids analyzer, it was calculated with a mass balance around the aeration basin as shown in Equation 4-6, which eliminated the need for one solids analyzer.

$$\frac{d(MLSS)}{dt} = \frac{(QiXi) + (QrRAS) - ((Qi+Qr)MLSS) + (RV)}{V} \quad [3]$$

[4-6]

where: R = solids growth rate ($M/L^3/T$)

Qi = influent flow rate (L^3/T)

Xi = influent solids concentration (M/L^3)

RAS = return activated sludge concentration (M/L^3)

Qr = recycle flow rate (L^3/T)

Equation 4-6 can easily be solved in real time for the value of MLSS. This *calculated* MLSS concentration based on the mass balance equation is then used in Equation 4-5 in place of a measured value. Terms such as the influent solids (Xi) and growth rate (R) were estimated as monthly average values but could have been completely ignored since they are small components in the equation and are continuously corrected with the measured value of RAS concentration.

An unexpected benefit of using the calculated MLSS concentration rather than a measured value was eliminating the need to accurately calibrate the solids analyzer for control purposes. This advantage can be seen by imagining that the RAS analyzer was measuring high by a factor of two. In this case, the calculated MLSS concentration would also be high by a factor of two. When these two values are plugged into Equation 4-5 to calculate the sludge wasting rate, they exactly cancel each other out. A more formal analysis of the problem showed that the only requirement was that the RAS instrument had to be linear with solids concentration. Most solids analyzers have this property.

The advantages of the Tillsonburg strategy:

- Operator attention, calculations, and adjustments were reduced.
- Laboratory analyses were reduced.
- Modeling eliminated the need for one solids analyzer (MLSS).
- Model-based strategy eliminated the need to accurately calibrate RAS analyzer.

Collingwood, Ontario, Canada

Automatic sludge age control was implemented at Collingwood, Ontario as a component to stabilize solids while different chemical feed strategies were evaluated for phosphorus control. The most unique aspect of the Collingwood implementation is that the plant used a steady-state mass balance around the settler to relate the RAS and MLSS concentrations to hydraulic flows. This mass balance is shown in Equation 4-7. The effluent solids were considered negligible in this mass balance. If effluent solids were not very low, they could not have been ignored in Equation 4-7.

$$Qr \times RAS = (Qi+Qr) \times MLSS \quad [4] \qquad [4\text{-}7]$$

The terms of this equation can be plugged into the formula for waste flow rate (Equation 2) to yield an equation for sludge wasting flow rate that depends only on hydraulic (flow rate) measurements and does not require any solids measurements. Note that Equation 4-7 may not be appropriate for overloaded clarifiers.

Collingwood was able to greatly improve its control of sludge age with the strategy shown in Equation 4-8.

$$Qw = \frac{V \times Qr}{SRT \times (Qi+Qr)} \quad [5] \qquad [4\text{-}8]$$

Calculated sludge ages prior to and after implementing automatic control are shown in Figures 4-8 and 4-9. The improvements are obvious.

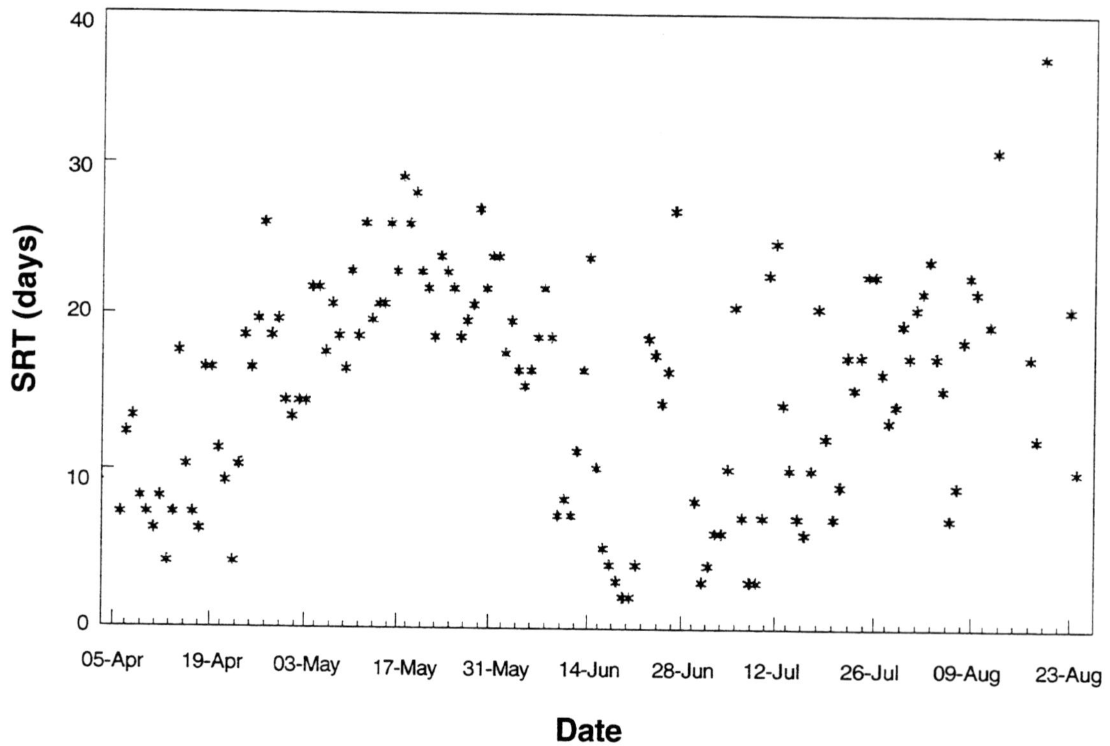

Figure 4-8. Collingwood WWTP Daily SRT Values Prior to Automatic SRT Control

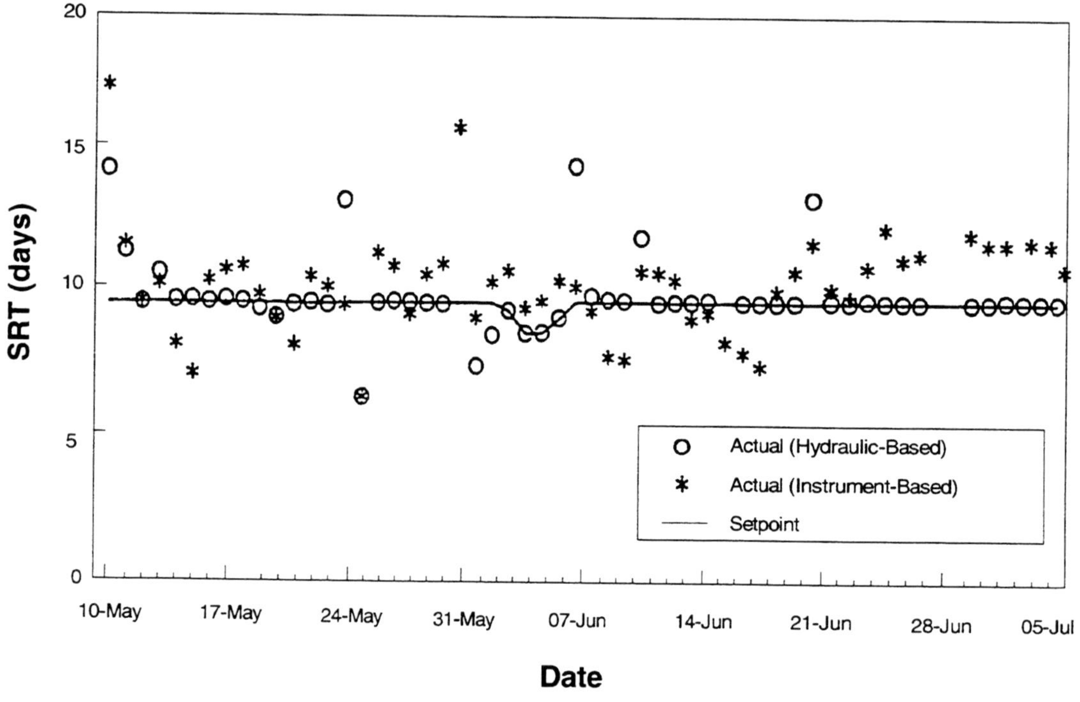

Figure 4-9. Collingwood WWTP Daily SRT Values with Automatic SRT Control

The advantages of the Collingwood strategy include:

- improved SRT control compared to the manual method previously used;
- reduced operator attention, calculations, and adjustments;
- reduced laboratory analyses; and
- elimination of the need for all solids analyzers.

Houston, TX

At many of its facilities, Houston practices a method of automatic sludge age control that was initially published in the literature by Dr. M. T. Garrett in 1958. The Garrett method utilizes wasting of MLSS directly from the aeration basin rather than the thicker RAS stream. In this way, the sludge age calculation is again reduced to a hydraulic problem as shown in Equation 4-9. If a sludge age of 10 days is desired, then $1/10^{th}$ of the aeration basin volume is wasted each day. This equation also assumes that effluent solids are insignificant.

$$Qw = \frac{V \times MLSS}{SRT \times MLSS} \quad [6]$$

[4-9]

The advantages of the Garrett method:

- Operator attention, calculations, and adjustments are reduced.
- Laboratory analyses are reduced.
- The need for all solids analyzers is eliminated.
- Only a single flow measurement (Qw) is required.

The primary disadvantage of the Garrett method is that the MLSS stream may need to be thickened before further treatment.

Madison, WI

The Madison facility has been rather successful in almost all phases of automating the plant with the exception of suspended solids analyzers. Since the Madison plant has the ability to waste either MLSS or RAS, it implemented two strategies that are essentially identical to the Collingwood and Garrett methods of hydraulic sludge age control, with a software switch to switch between the two strategies.

4.6.4.4 Conclusions on Automatic Sludge Wasting Control

1. A number of innovative strategies exist for automatic control of sludge age.
2. Demonstrated benefits of automatic sludge age control include improved process performance, less operator attention and adjustment, decreased number of laboratory analyses, reduction of variability to downstream processes, and improved sludge settling characteristics.
3. Control strategies have been proven on numerous facilities.
4. Solids analyzers have proven effective in some but not all facilities.

4.6.5 Automatic Control of Alum Addition for Phosphorus Removal – Collingwood, Ontario

Imposition of stringent phosphorus limits (0.31 mg/L) required close control of alum addition rates for precipitation of soluble phosphorus in the activated sludge process. To achieve the target performance, Collingwood implemented an automated chemical dosage control, along with automated SRT control (Brewer et al. 1996). Key elements of the monitoring and control system included:

- two on-line colorimetric phosphate (PO_4) analyzers;
- sample delivery systems to pump samples from the primary effluent channel feeding the aeration tanks or from the mixed liquor channel feeding the secondary clarifiers to sample conditioning equipment;
- sample conditioning systems consisting of ultrafilters;
- permeate pumps to deliver the ultrafilter permeate to the phosphate analyzers;
- a PLC to monitor the analyzer output and control the speed of the alum feed pump;
- alum metering pumps with variable speed capabilities and flow meters; and
- a PC-based supervisory control and data acquisition system.

The control strategy involved a closed control loop in which alum was added to maintain the primary effluent or the aeration tank effluent phosphorus concentration between preset limits, depending on the location of alum addition. If alum was being added to the primary clarifier influent (pre-precipitation), a setpoint of 2.0 mg/L was established for the primary clarifier effluent phosphorus concentration. If alum was being added to the aeration basin, a setpoint of 0.1 mg/L phosphorus was established for the mixed liquor entering the secondary clarifiers. The capability to flow proportion alum addition in response to raw sewage or final effluent flow meters was also provided. A P&ID illustrating the control strategy is presented in Figure 4-10.

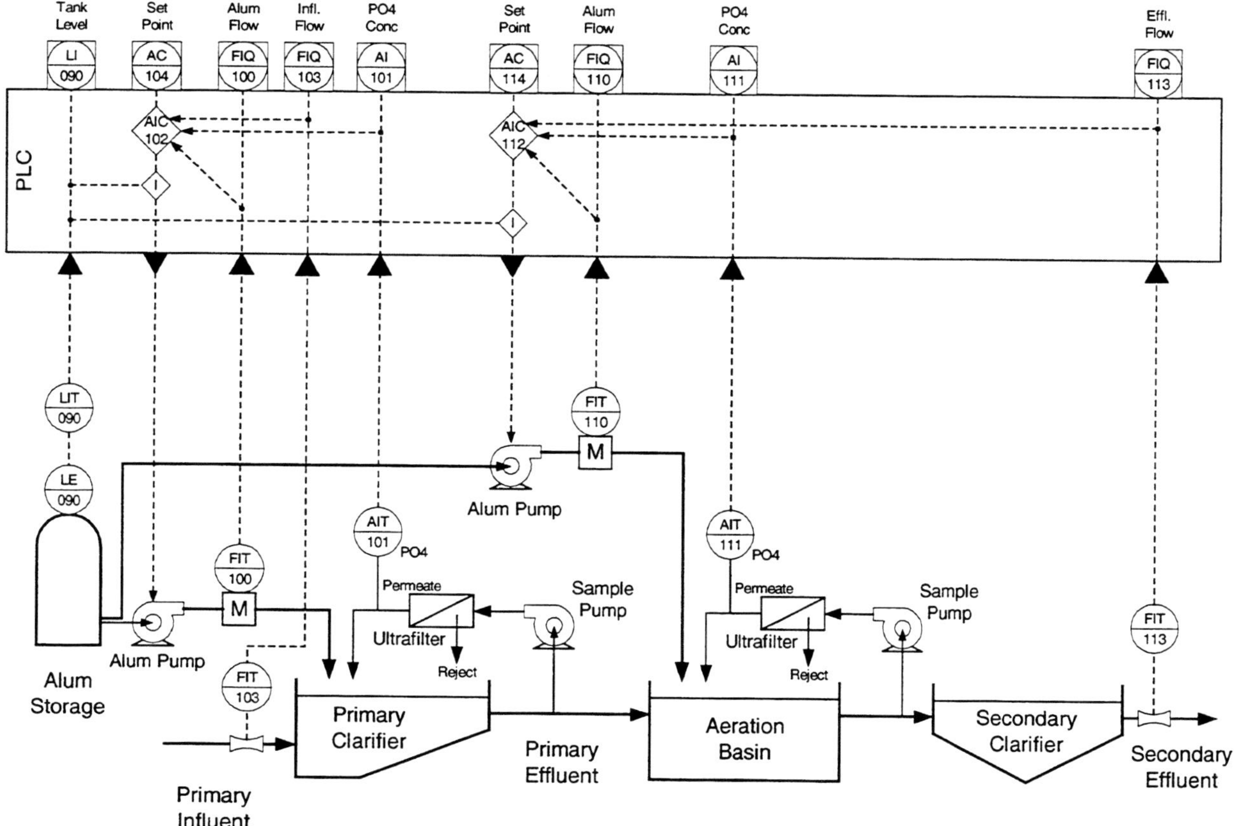

Figure 4-10. Collingwood WWTP P&ID for Alum Addition for Phosphorus Control

As a result of the automation of alum addition, plus automated control of SRT, the facility has been able to achieve a final effluent total phosphorus concentration consistently below the target level of 0.31 mg/L when the control system was operated on either primary effluent or mixed liquor. Figures 4-11 and 4-12 show the response of the alum feed system to changes in ortho-phosphate concentration. At the same time, an overall reduction in alum consumption was also noted as a result of the on-line monitoring and control system. Maintenance of the control system was primarily associated with cleaning of the ultrafiltration membranes, permeate piping, and analyzer cell. The estimated annual cost to operate the automated chemical addition system is $20,000 (1996 dollars). Of these costs, about 66% were associated with operational labor.

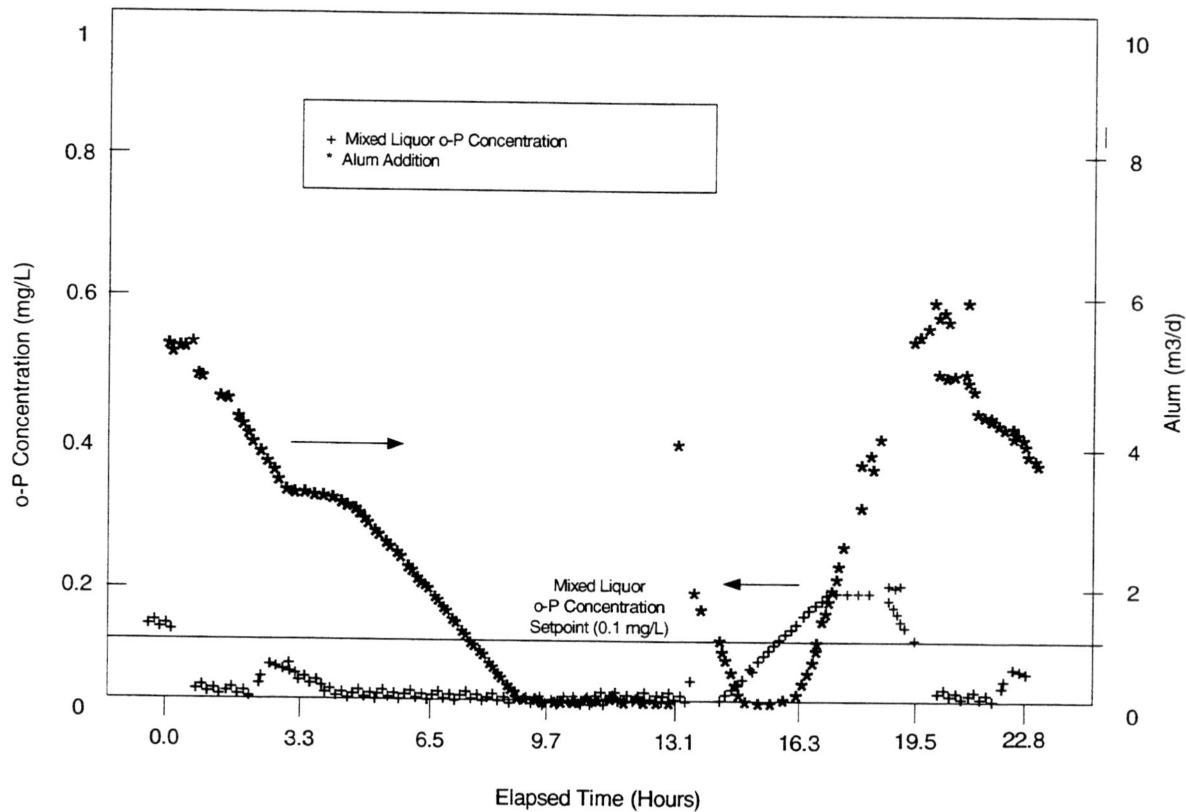

Figure 4-11. Collingwood WWTP Alum Addition to Deviation of Control Variable

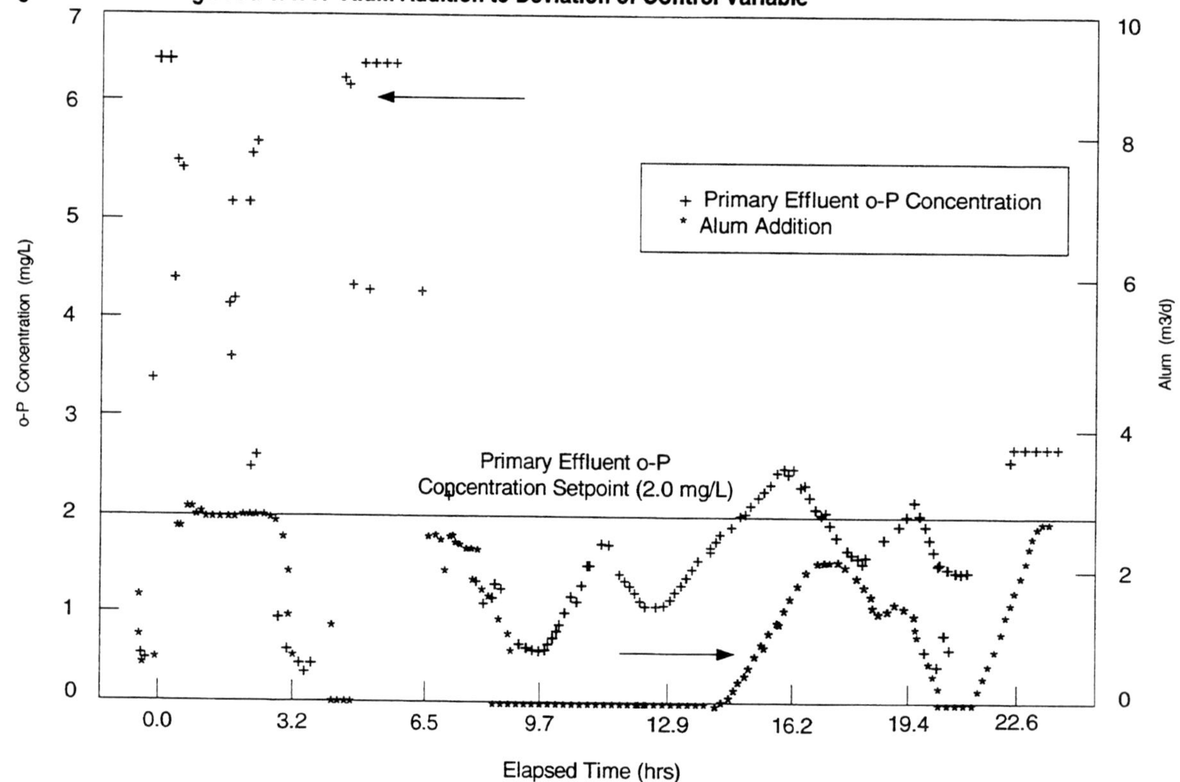

Figure 4-12. Collingwood WWTP Alum Addition to Deviation of Control Variable

4.7 Solids Processing (Thickening, Stabilization Dewatering)

4.7.1 Waste Activated Sludge Thickening Control – Collingwood, Ontario

Prior to implementing automatic SRT control (refer to Section 4.6.4.3), waste activated sludge (WAS) from the secondary clarifiers was thickened in a dissolved air flotation (DAF) thickener. WAS feed was provided to the thickener by a manually adjustable variable speed pump. Performance of the DAF thickener was very sensitive to both hydraulic loading and solids mass loading. Poor thickener performance had an adverse affect on the downstream anaerobic digestion process and high polymer dosages were used in an attempt to maintain thickener performance.

When the automatic SRT control system was implemented, the manually adjustable variable speed control on the DAF thickener feed pump was replaced with an automatic speed controller. In conjunction with a magnetic flowmeter and suspended solids monitor on the WAS stream (both components of the SRT control system), the feed rate to the DAF thickener could be optimized based on maximum hydraulic and solids loading rates calculated by the SCADA software from operator set-point values. Automatically actuated valves were installed to allow any WAS beyond the capacity of the DAF thickener to be diverted to the primary clarifiers to be co-thickened. A P&ID illustrating the control system is presented in Figure 4-13.

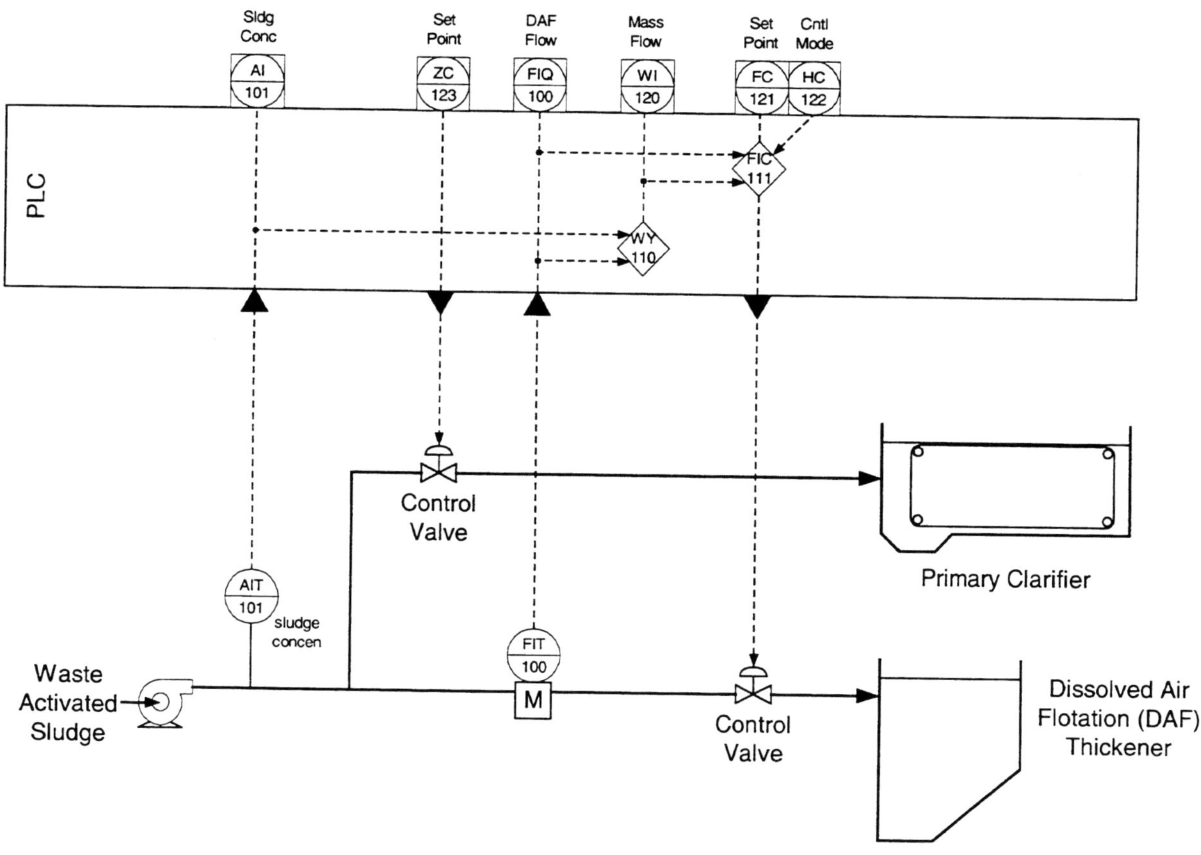

Figure 4-13. Collingwood WWTP P&ID for DAF Control

Since implementation, the performance of the DAF thickener has stabilized, reducing polymer use and improving digester performance. Costs of implementation were minimal as the sensors and control hardware were available to support the automated SRT control system.

4.7.2 Sludge Blanket Control of Gravity Thickening – Houston, TX

The City of Houston's 69th Street wastewater treatment facility treats 3.29 cubic meters per seconds (75 mgd) of wastewater with a pure oxygen activated sludge process without primary sedimentation. The plant treats its own waste activated sludge and sludge from several satellite facilities for a total of over 90,000 dry kilograms (100 tons) per day. Sludge treatment consists of fine screening, gravity thickening, aerobic digestion, dewatering by centrifuges, and flash drying. The final dry product is marketed as a soil additive and low-grade organic fertilizer. The gravity thickeners may be used to pre-thicken the sludge before digestion, post-thicken the sludge after digestion, or pre-thicken the sludge and send it directly to dewatering. The following strategy discusses the latter mode and was used for a number of years.

Polymer costs for sludge dewatering were relatively high because of the dilute nature of the aerobically digested sludge. Plant staff observed that when solids concentrations to the centrifuges were high and held relatively constant, polymer costs were greatly reduced.

Plant staff decided to automate one gravity thickener to achieve maximum solids concentration. Since compression in the thickener is a function of detention time and mechanical stress, they reasoned that the maximum concentration would be achieved by maintaining as high a sludge blanket in the thickener as possible. The danger of this strategy is that if the sludge is held too long, gas bubbles form that virtually destroy the thickening and lead to high-suspended solids in the supernatant. Through a process of trial and error, the staff found a suitable sludge blanket level that achieved concentrated thickened sludge concentrations without other adverse effects.

Figure 4-14 is a P&ID of the strategy adapted for thickener control. A sludge blanket interface instrument was used to measure the sludge blanket level in the thickener. The instrument was based on an optical suspended solids probe mounted on a reel that periodically lowered the probe down until it detected the sludge blanket level. Numerous enhancements were made on this instrument to improve its reliability and reduce maintenance. The blanket level was then used as the process variable for a very slow-acting control loop that modulated the underflow pumps.

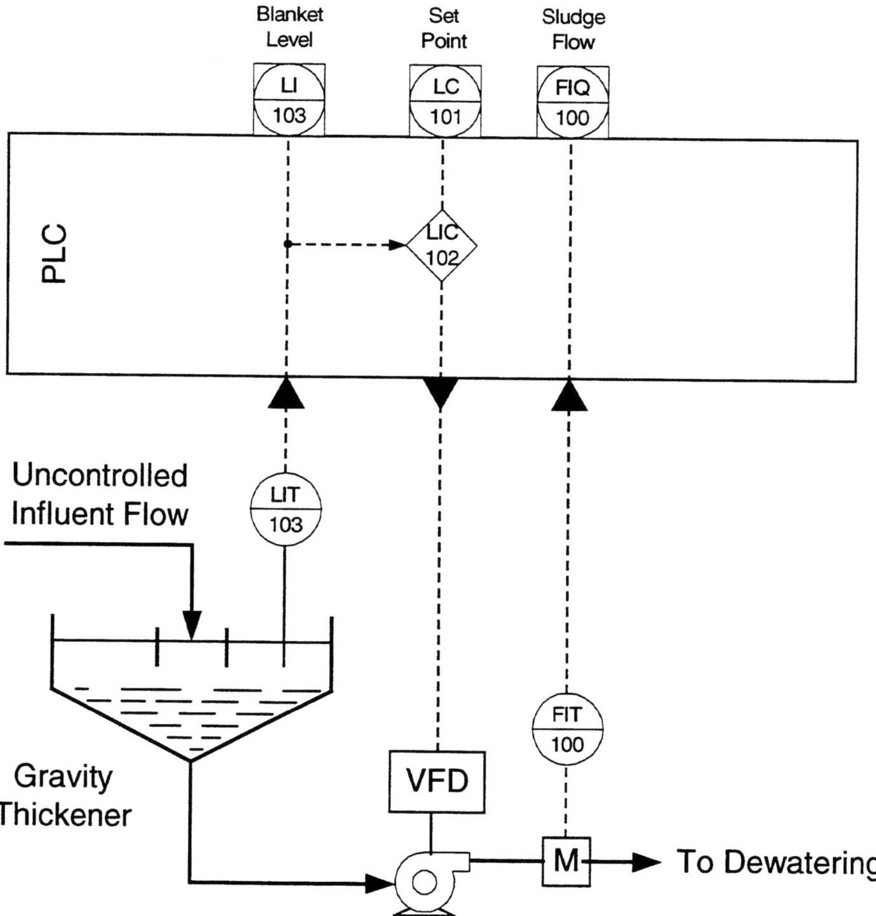

Figure 4-14. P&ID for Houston's Control of Gravity Thickening

The thickener automation was very successful. Sludge concentrations were increased from about 3% solids with much variation to nearly 4% solids and relatively constant. Polymer costs were reduced 47% with a resulting savings of several hundred thousand dollars per year.

4.7.3 Automatic Sludge Conditioning Control – Durham Region, Ontario

Efforts to optimize sludge conditioning to improve sludge cake release and cake dryness from sludge filter presses prompted plant staff to install an automated polymer feed control system (Kent, 2001). The proprietary control system provided measures sludge density, sludge and polymer feed rates, and pressure within the press. Sludge feed rate, polymer feed rate, and mixing valve position are continuously adjusted to maintain optimum filter press operating conditions. A P&ID illustrating the control strategy is presented in Figure 4-15.

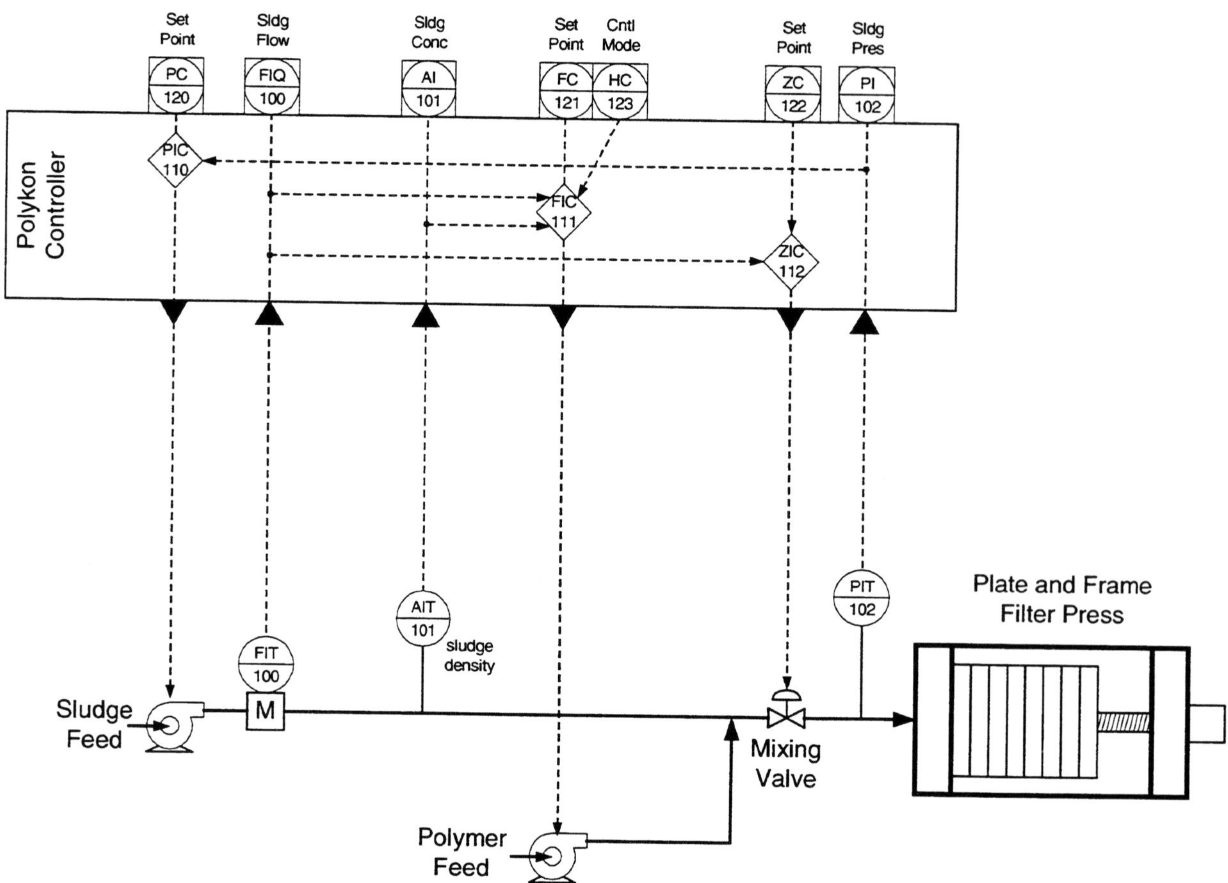

Figure 4-15. P&ID for Automated Control of Solids Dewatering

Polymer dosage can be regulated based either on a volumetric dose (g of polymer per m^3 of sludge feed) or a mass dose (kg of polymer per dry tonne of sludge feed). The press operator selects the control strategy used and the dosage to be applied, provided as the control setpoint to the controller. When operated in the volumetric-proportional mode, the polymer feed rate is adjusted based on the output signal from the sludge magnetic induction flowmeter. This control strategy is effective when the feed sludge concentration is relatively constant. Mass-proportional dosing ensures that the polymer dosage per unit mass of solids fed to the press is maintained constant despite fluctuations in flow and sludge concentration. In this operating mode, sludge concentration is measured using a nuclear density meter and the mass in dry solids/hr is calculated. The polymer feed pump is then adjusted to maintain a constant mass dosage rate of polymer.

Throughout the filling cycle, the mixing valve is adjusted to maintain ideal mixing of the polymer and sludge as the sludge flow rate declines with time through the press cycle. At the same time, the press pressure is used to adjust the sludge feed rate to maintain the ideal setpoint.

Since implementation of the polymer conditioning control system, the plant has increased the sludge throughput of its presses by about 25% because of better cake release from the presses. At the same time, cake dryness has increased marginally with a slight increase in polymer use. Lower maintenance costs, less down time, and lower fuel usage in the sludge incinerators have resulted in a payback of about 2.8 years on the control system.

4.7.4 Unstaffed Startup of Gravity Belt Press – North Shore Sanitation District, IL

Rethinking operations to move from 24/7 staffing to weekday staffing required review of process operations to determine which process activities require supervision compared to the process activities that have typically been supervised but can be modified to run unattended, or partially attended. The North Shore Regional Sanitary Sewerage District reported unattended, automated startup of their Gravity Belt Press operation at 3:00 a.m., prior to staff arrival at the facility for the day shift. The staff is required to handle the thickened sludge, but the thickening process starts up and runs automatically and unattended.

At the time of the interview, the sludge blending operation was the least predictable, and the plant staff was considering options for improving automatic polymer addition, including vision systems (video cameras) and changing the polymer addition location.

4.7.5 Automated Supernatant Return – Denver, CO

Anaerobic digester sludge has a high ammonia concentration and can be a significant load to a nitrifying activated sludge facility. At the Metro Wastewater Reclamation District, the ammonia load from centrifuge centrate was about 50% of the influent load. When the facility was upgraded for nitrification, control of this load was a prime design consideration. A centrate storage tank was designed so that the centrate could be fed back into the plant at a rate that would be economical and would not upset the process.

Shown in Figure 4-16, the control strategy for feeding centrate has four possible control modes: level control, flow rate control, ratio to total secondary influent flow, and manual. The operators normally run the process with level control where the flow rate is proportional to the level of centrate in the storage tank.

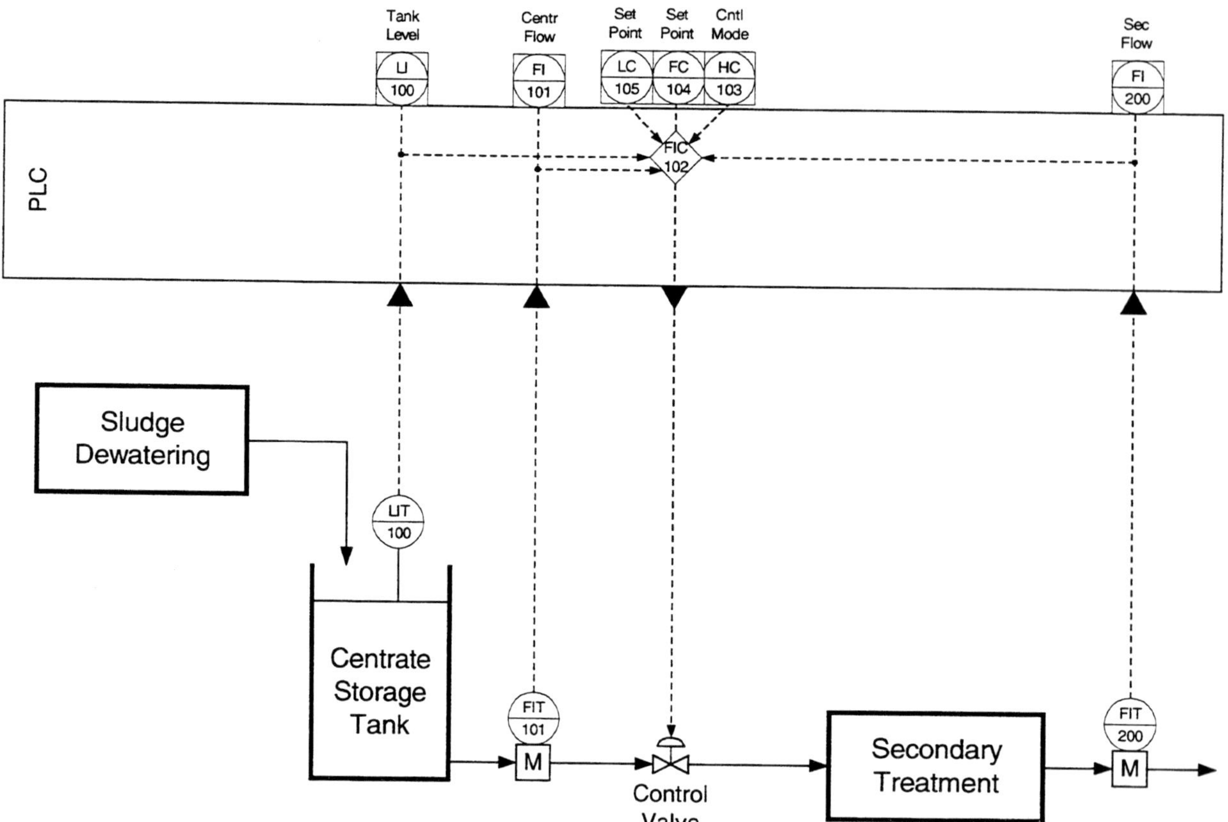

Figure 4-16. P&ID for Centrate Feed Controller

The actual control strategy in practice is somewhat more complicated than that shown in Figure 4-16 because it also included four pumps not shown on the diagram. This complication was not necessary to show the concept.

4.8 Effluent and Disinfection Processes

4.8.1 Flow Paced Ultraviolet Disinfection

Ultraviolet (UV) disinfection is a non-chemical disinfection alternative that uses low-pressure mercury lamps that are believed to break the DNA and RNA chemical bonds, thereby blocking cell replication. The mercury lamps produce UV light in the region of 250-265 nanometers, critical for cell destruction.

The UV lamps are configured in banks, creating an array of tubes that the effluent wastewater flows through. The bank of lamps may be set either perpendicular or parallel to the channel water surface, depending on the type of system. Wastewater, as it passes close to a powered lamp, is disinfected. The UV lamps are typically installed in an effluent channel, and the required contact time with an active lamp is typically 6-8 seconds.

UV disinfection control systems are typically bundled with the entire UV process. Several options are available for UV lamp control. Selection of the controls depends on the process, the types of lamps in use, and the wastewater characteristics, among other things.

The disinfection of final plant effluent utilizing UV light employs a feed forward control system, which consists of a series of lamps and or lamp channels. The number of lamps turned on at any time is proportional to the plant flow. The sequencing strategy typically controls the lamp on-time and off-time to prevent cycling lamps and to optimize lamp life. The typical daily flow pattern for the plant can be used to determine the base load of lamps required. The actual plant flow can then be used to trim the number of lamps running.

The effectiveness of UV disinfection depends on lamp intensity, which is directly affected by turbidity and by lamp fouling. Regular maintenance of the lamps includes frequent cleaning of the tubes. Control strategies that set the number of lamps in operation should take these operating conditions into account.

The result of the bacterial kill analysis typically takes about 24 hours, making a feedback control measurement impractical. UV transmittance analyzers may be utilized for monitoring system performance but are not generally employed in feedback control.

4.8.2 Dechlorination When Outfall Provides Contact Time - Physical Model Simulation – MWRA, Boston, MA

Current chlorine discharge permits require a 15-minute chlorine residual measurement and dechlorination to prevent discharge of chlorine residual into receiving waters. If the plant effluent were to reach its 15-minute contact in a tunnel 400 feet below the surface of the ocean, possible solutions include adding a sample line and pumping sample back to a sampler. However, that significantly increases the process dead time, so the measurement isn't a true 15-minute residual.

Due to the long distance and volume in the outfall tunnel, the contact time does exist in the process. In fact, at normal plant flows, the 15-minute residual can be measured in the contact chambers. But at high plant flows, the 15-minute point is located in the outfall tunnel. The permit allows for controls to use a simulation loop, operated in a similar fashion to the process, to provide a residual measurement representative of the residual in the outfall tunnel. Another strategy uses a plant model designed to treat plant effluent in parallel to the actual process treatment. Measurements made in the model loop are representative of the real process and so eliminate the need to try to sample the effluent in an inaccessible location.

The model loop includes instruments and process similar to the actual outfall. Chlorine residual is measured in the model system, similar to the measurement at the end of the contact basin. Sodium bisulfite is injected at an equivalent injection location and resulting chlorine residual measured to determine the effectiveness of the control.

To compensate for a residual measurement that varies with flow, and to provide reliability in the controls, several different control methods are available for calculating the appropriate sodium bisulfite flow rate. The operator may select the mode of operation.

4.8.2.1 Basin Mode

Bisulfite is added at the end of the disinfection basin downstream from the chlorine residual analyzer. In basin mode the sodium bisulfite flow setpoint is calculated as a function of the measured chorine residual compared to the required residual and corrected by the sodium bisulfite concentration (Figure 4-17).

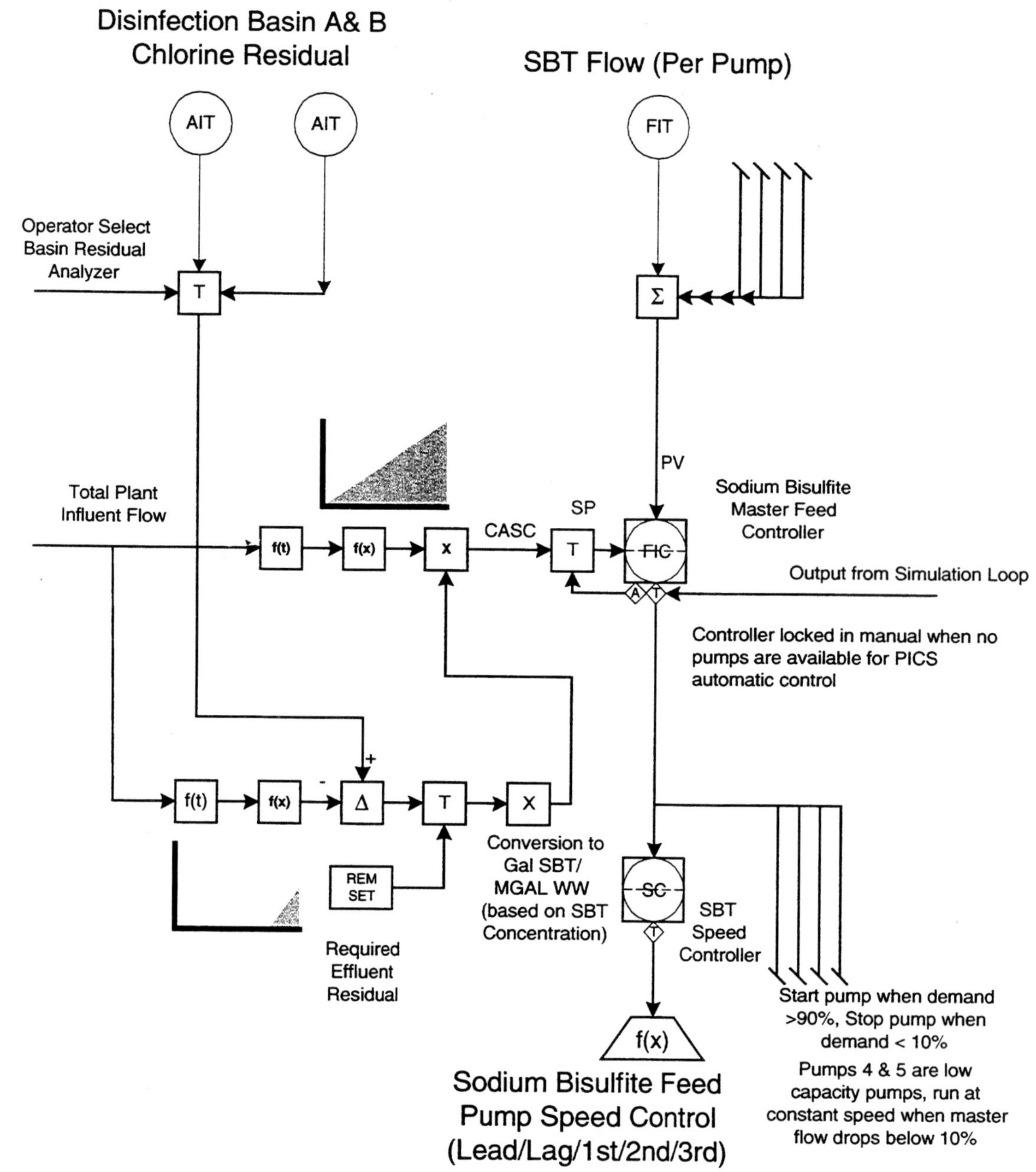

Disinfection Basin A& B
Chlorine Residual

SBT Flow (Per Pump)

AIT
AIT
FIT

Operator Select
Basin Residual
Analyzer

T
Σ

PV

Total Plant
Influent Flow

f(t)
f(x)
x
CASC
SP
T
FIC

Sodium Bisulfite
Master Feed
Controller

Output from Simulation Loop

A
T

Controller locked in manual when no
pumps are available for PICS
automatic control

+
-
f(t)
f(x)
Δ
T
X

REM
SET

Conversion to
Gal SBT/
MGAL WW
(based on SBT
Concentration)

SC
T

SBT
Speed
Controller

Required
Effluent
Residual

Start pump when demand
>90%, Stop pump when
demand < 10%

f(x)

Sodium Bisulfite Feed
Pump Speed Control
(Lead/Lag/1st/2nd/3rd)

Pumps 4 & 5 are low
capacity pumps, run at
constant speed when master
flow drops below 10%

Figure 4-17. Sodium Bisulfite Feed Control

4.8.2.2 Simulation Loop Mode

In simulation mode, a sample process loop runs in parallel with the sodium bisulfite loop. The chlorine residual concentration required to achieve permit limits for chlorine residual in the effluent is calculated based on the process flow through the simulation loop. The calculated residual is compared to a residual measured at that location in the simulation loop. (The operator may select one of two different sample locations; the required effluent is calculated for both sample locations but only one calculation is used.)

The difference between the calculated and measured residual is converted to a required sodium bisulfite rate per unit of plant flow. The rate is multiplied by the actual plant flow, to determine the setpoint for the sodium bisulfite feed master, and the simulation feed master (Figure 4-18).

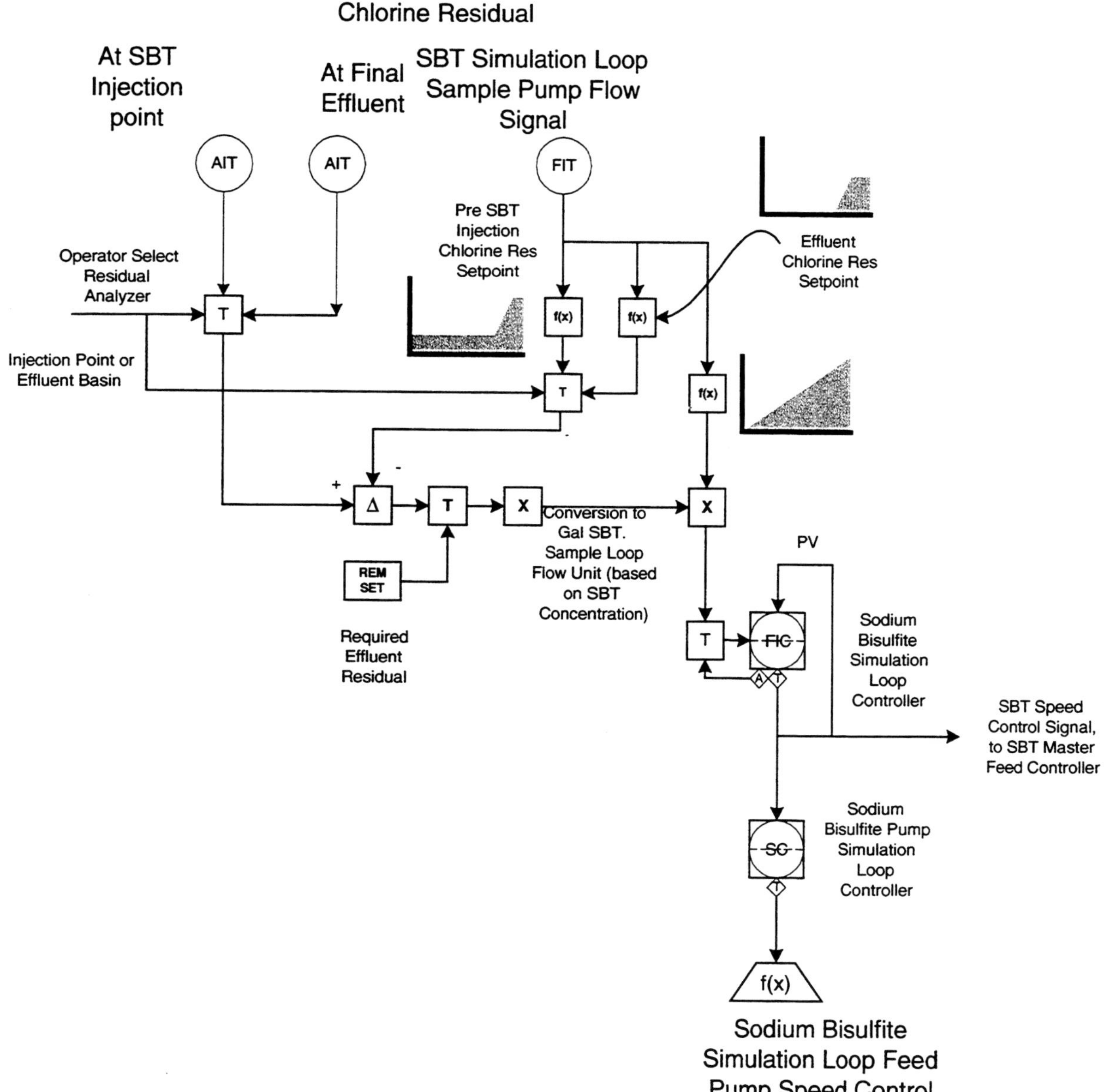

Figure 4-18. Sodium Bisulfite Simulation Loop Control

4.9 Primary Sensors (Instrumentation)

Rapid advances in control hardware and software now allow the smallest wastewater treatment plants to economically implement automation. The control hardware is relatively inexpensive and simple to install, program, and maintain.

Even the simplest control strategy depends on an accurate signal from the primary sensor. Significant development in the field of sensor technology has occurred over the past two decades. A recent WERF review (ITA, 1994) found that on-line instruments were commercially available or under development for nearly any analytical methodology available in the laboratory. Despite this development, the dependability and accuracy of the primary sensors is still cited in the literature, in automation workshops, and among practitioners as the single largest impediment to wide-scale, successful implementation of automation. Participants at the 2000 workshop conducted as part of this project confirmed that sensor accuracy and reliability continues to be a problem area.

The problems associated with field sensors are not unique to the wastewater treatment industry. Sanders (1995) reported on a chemical process plant that found that two thirds of its upsets could be traced to instrument faults. As noted by McMillan et al. (1998), improperly selected and maintained instrumentation can increase maintenance costs as well as reducing plant operating efficiency and capacity. McMillan et al. (1998) summarizes some of the most common causes of measurement errors and failures in the chemical process industry:

- sensing lines that are plugged or that contain liquid when they should be dry;
- sensing elements with excessive coating, fouling or abrasion;
- excessive bubbles or solids in process fluids;
- sensing elements with deformations, cracks and holes;
- low Reynolds number for process fluid;
- gaskets or O-rings that leak;
- inappropriate materials of construction;
- sensing, pneumatic or electronic components affected by process or ambient temperature;
- moisture on the sensing element or signal connections;
- electrical interference;
- high connection or wiring resistance;
- non-representative sampling point;
- inadequate straight-pipe runs for flow sensor;
- nozzle flappers that are plugged or fouled;
- feedback linkages that shift or contain excessive play; and
- incorrect calibrations.

The causes of these sensor errors could equally well have been cited for the wastewater treatment industry, a particularly harsh environment for instrumentation. The first three causes have been found to be particularly problematic in the wastewater industry.

In 1977, a U.S. EPA-sponsored survey of 50 wastewater treatment plants found that 31% of the process control instruments had a performance record that was less than satisfactory (Roesler, 1977). Despite manuals of practice and guidance documents aimed at improving

instrumentation performance and reliability (WPCF, 1978; WEF, 1993), the record today may not be significantly better than it was in 1977.

Adopting good engineering practices for instrumentation selection, installation, and maintenance can improve the reliability and dependability of primary sensors, which is essential if operations staff are to have confidence in the overall control system. These practices must also be applied to the design of any sample transport and conditioning system that may be associated with the sensor.

4.9.1 Components of a Sensor

Accurate and reliable on-line measurements depend on the successful operation of all components of the sensing system, which may, depending on the variable being measured, include some or all of the following: sample transport, sample conditioning, measurement, signal conditioning, and signal transmittal.

A typical sensor installation showing these components is presented in Figure 4-19. Extensive testing by agencies such as the ITA has shown that most sensors are capable of accurately measuring the parameter of interest and transmitting a reliable signal under ideal conditions in the laboratory (ITA, 1988). Sensor reliability problems are more often related to sample transport, sample conditioning, and sensor fouling under field operating conditions than poor analytical methodology or inferior electronics. Interestingly, these problems are the first three identified by McMillan et al. in their review of instrumentation problems in the chemical process industry.

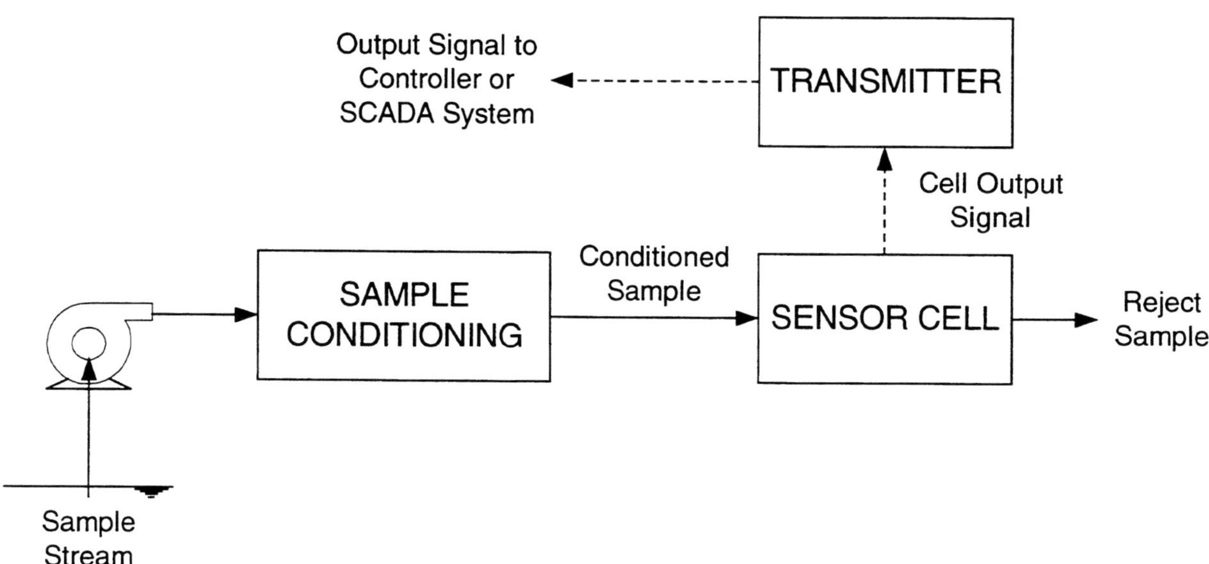

Figure 4-19. Block Diagram of a Sensor System

4.9.1.1 Sample Transport

Some sensor technologies, including some analyzers that measure parameters such as residual chlorine, ammonia, phosphorus, and turbidity, are not capable of making an in-situ measurement at the point of interest. These sensors need a sample to be delivered from the measurement point to the sensor's measurement cell. A sample transport loop, typically comprising a sample pump, sample transport piping, and a spent sample disposal system, must be installed. The requirement for sample transport adds complexity to the instrumentation system with additional equipment that requires maintenance and cleaning. A poorly designed or poorly maintained sample transport system can render the most accurate and reliable sensor useless. McMillan et al. (1998) conclude that sample transport lines are the biggest source of maintenance problems in instrumentation systems.

To design an effective sample transport system:

1. Avoid sample transport if possible. Where necessary, minimize the sample transport distance as much as possible. Long sample transport lines cause significant lag time in a control loop if the measured variable is used for control purposes. Non-conservative parameters, like ammonia and dissolved oxygen, can change in concentration during transport from the sample point to the measurement point. Short transport lines also reduce maintenance requirements and chances of plugging.

2. Size the sample pump to deliver the required sample volume to the sensor under all conditions. Depending on the nature of the sample, grinder pumps may be required to prevent plugging of sample lines. Pre-screening of the sample could also be considered, but steps must be taken to prevent the screen from plugging. Oversizing of the sample pump and bypassing the excess flow at the sensor will reduce transport time and minimize the chances of plugging in small sample lines.

3. Ensure that the sample location is representative of the stream being analyzed. Locate the suction of the sample pump in a well-mixed area. Verify through manual sampling and off-line analysis that the sample location is representative. Also, verify that the parameter being measured does not change in the sample line during transport to the analyzer.

4.9.1.2 Sample Conditioning

Sample conditioning can refer to either:

- addition of reagents to the sample stream to allow the measurement of the analyte in the measurement cell (for example, a colorimetric analyzer in which reagents are added to the sample to react with the chemical being measured and produce color); or,

- treating the sample to remove constituents that can interfere with the measurement or foul the sensor (for example, filtration to remove particulate matter).

Reagent addition is normally an intrinsic part of the sensor and the reagent pumps; piping and other components are part of the analyzer design. The design and operation of these components should be reviewed as part of the sensor selection process.

Sample treatment, by filtration or other means prior to analysis, can represent a significant problem due to the nature of the samples being analyzed in wastewater treatment plants. Some manufacturers provide sample conditioning equipment, such as cartridge filters, Y-

strainers, screens, and settling tubes, as an accessory for their sensors. Many wastewater treatment plant operators have designed and installed their own sample conditioning equipment because of the poor reliability of the equipment provided by the manufacturers. At the Collingwood WWTP, a membrane filtration system with automatic cleaning capability has been successfully used to condition both primary effluent and mixed liquor prior to a continuous colorimetric phosphorus analyzer (CH2M Hill Gore & Storrie, 1996). This system is described in more detail in Section 4.6.5.

ITA (1994) identified sample transport and conditioning systems as the prime cause of sensor failure in wastewater monitoring applications. This was borne out by more recent ITA testing with eight on-line ammonia analyzers (ITA, 2001). Further, the WERF research needs study found that there was virtually no ongoing research activity and little published information on the subject of sample transport and conditioning. ITA, with wastewater utility support, conducted an initial study of sample conditioning systems that provides very beneficial information in this problem area (ITA, 2001). The full study is waiting for additional funding.

4.9.1.3 Sensor Selection

Because the best control system cannot function without accurate, reliable primary sensors, selection of the most appropriate sensor for a specific application is critical to success. Most sensors perform well under the ideal conditions the manufacturers use to determine their specifications for accuracy, reproducibility, etc. However, sensor performance in the field is sometimes less than satisfactory.

Field testing of sensors is invaluable in providing the information needed to make informed sensor procurement decisions. Such testing can evaluate the entire sensor system, including any sample transport and conditioning components, and provide information on the cost of ownership. Although individual testing in the plant by the end user is the best way to select the best sensor system for a particular application, such testing can be costly and time consuming. Hence, testing organizations have been established that conduct standardized instrument tests against rigorous test protocols.

A number of such testing organizations are located in North America and in Europe (ITA, 2000). All have a similar mission – to provide performance evaluation information to the users of instrumentation. The test reports produced by these organizations can be used to evaluate sensor performance and assist with sensor selection.

4.9.2 Sensor Maintenance

According to McMillan et al. (1998), instrument maintenance has become one of the largest categories of operating cost in chemical process industry (CPI) plants. It is doubtful if instrument maintenance receives this level of attention in even the most highly automated wastewater treatment plant, which may help explain the lack of acceptance of automation in the industry and the high number of cases where plant staff report that their automation system "doesn't work." Modern CPI plants are automated to the extent that the operation of the process control system is essential to production and profitability. Such is not the case in wastewater treatment plants where "operating in manual" remains a relatively simple and viable option.

The level of maintenance needed to ensure reliable and accurate instrument operation depends on the type of sensors. It can also be influenced by the design of the sensor installation. Poorly designed or inadequate sample conditioning equipment can result in a high level of maintenance, frustration among the operating staff, and, in the worst cases, complete abandonment of the sensor and the loss of process control. ITA has published maintenance benchmarking studies for a variety of sensors, namely total and free chlorine residual analyzers, pH analyzers, suspended solids analyzers, and turbidity analyzers (ITA, 1999) and is planning additional surveys (ITA, 2000). These reports provide guidance on the frequency and duration of maintenance applied to these sensors by wastewater industry users as well as information on the costs and effectiveness of the maintenance programs applied.

McMillan et al. (1998) acknowledge that, in CPI plants, many requested instrument maintenance and calibration checks prove unnecessary. For example, in one large chemical plant, 35% of instrument checks made for preventative maintenance and 28% of instrument checks for reactive maintenance found no problem (Rosemount, 1998). Finding the proper balance of maintenance to ensure reliability and accuracy in the sensor signal while minimizing costs and eliminating unnecessary maintenance is the critical to the success of an automation system.

Computerized maintenance management systems (CMMS) can be valuable tools in optimizing instrument maintenance. Many of the wastewater utilities included in ITA's maintenance benchmarking studies reported using CMMS to schedule and record instrumentation and process control system maintenance. Information contained in the CMMS can provide guidance on the frequency of maintenance actually needed to ensure proper operation of the instrumentation.

Smart sensors, equipped with self-diagnostics and auto-calibration features, have become common, and instrument manufacturers continue to develop new diagnostic checks. For example, changes in signal-to-noise amplitude can be used to detect plugged, coated, or stuck sensors, or changes in response time in a closed-loop test can detect a fouled sensor (McMillan et al., 1998). Such sensors can reduce the overall cost of ownership by reducing unnecessary maintenance and calibration activities.

Calibration represents the largest component of instrument maintenance cost. According to McMillan et al. (1998), two-thirds of all calibration checks are unwarranted. Furthermore, the requirement for frequent recalibration of a sensor often is a symptom of a larger problem that has not been addressed.

Determining the need for instrument calibration must be based on the application of some consistent criteria. Quality control charts that compare sensor readings with an independent measurement of known accuracy have been shown to be an effective means of determining when calibration is needed. ITA successfully used this approach as part of its testing protocol for on-line sensors. Figure 4-20 shows the comparison between the output of an on-line DO sensor and a reference sensor (a calibrated handheld DO probe). A consistent deviation between the on-line sensor and the reference measurement indicates a need to clean or recalibrate the on-line sensor. A consistent deviation is defined as three consecutive measurement differences between the on-line analyzer and the reference measurement that exceed a pre-determined value (for example, 0.3 mg/L), and the difference is always in the same direction (ITA, 1988).

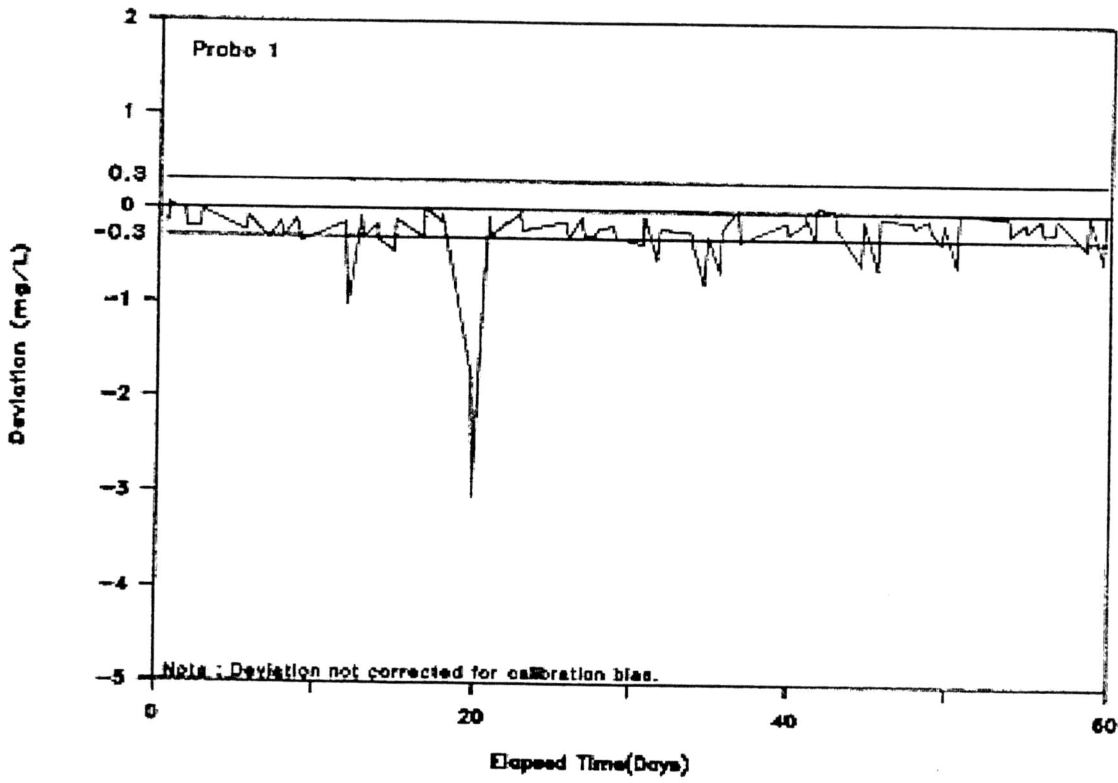

Figure 4-20. Example Test Result from an ITA DO Analyzer Test Report

(Copyright 1988, Instrumentation Testing Association, reprinted with permission, PER88DO-003, Performance Evaluation Report)

The analyzer is only recalibrated when cleaning does not eliminate the deviation between the on-line analyzer and the reference standard.

4.10 Control Hardware and Software

Throughout both the field surveys and the WEFTEC 2000 workshop, the issues identifying successful vs. problematic control systems tended to be less centered on the technology than on the methodology of implementation. In general, plant owners tended to be able to "make systems work" once installed and didn't generally undertake the effort and expense of replacing a process control system until such time as spare parts and service personnel were no longer available.

The steps to making each system work and become successful were investigated as part of the field survey. Each utility developed its own methodology for implementing and installing systems. The larger utilities with many treatment facilities typically had better developed procedures for the implementation and installation phases of a project.

4.10.1 System Selection/Procurement

The system selection process varies from facility to facility, based largely on procurement requirements of each utility. The relative cost of procurement was beyond the scope of this project; but the handicap of complex and restrictive procurement practices, though established with the good intention of ensuring a competitive procurement process, may offset the cost benefits by costing the owner more both for staff time and schedule impact.

At one utility (North Shore Regional Sanitary District) the executive director made the decision to cease hiring plant operators, and empowered the control system engineering group to automate all the facilities. This directive from management to implement process automation without any other restrictions served as a mandate to get the project started. The engineering group was able to proceed with equipment selection and procurement based on technical evaluation. The subsequent procurements allowed future sole source purchase of similar equipment, so that systems would consist of the same hardware and software.

The alternative to this practice is exemplified by several large treatment facilities that had process construction and upgrades built as separate construction packages. These large utilities purchased the local controls for the facility as part of the construction and established general specifications, including "three or equal" language when identifying control equipment vendors. "Three or equal" refers to specifications written to allow any one of at least three manufacturers to provide equipment. This type of procurement resulted in a hodgepodge of instruments and controls, including 12 different manufacturers of PLCs at one facility. This is obviously an extreme case and might be better controlled by tighter specification and more stringent review during the submittal process, but it highlights the worst-case scenario.

Most facilities described selecting equipment based on a technical evaluation, usually involving some combination of a written survey, an interview and product demonstration with the product manufacturer, calling a list of references, and site visits to facilities currently using the system. The combination of the technical evaluation coupled with a separate review of pricing established the vendor to be used for all facility upgrades.

4.10.2 System Implementation Including Testing and Startup

The steps taken to system implementation were critical to the overall success and to limit time during commissioning until the system was ready for startup. In general, the amount of owner involvement correlated directly with the degree of the system's success.

4.10.2.1 Design

In one case, the owner (NSRSD) described sending technical staff to work at the engineering offices of the firm preparing design documents. Alternatively, owners set aside office space for the engineering team and had them work right at the facility. This approach is one of the best ways to ensure constant communication between the design team and the owner and to make sure that the scope of work doesn't diverge in the minds of the two parties.

4.10.2.2 Implementation and Factory Testing

Once a system integrator or vendor has been selected, the plant owner may send engineers to work with the integrator during the system programming and configuration stages. Again, this procedure is an excellent, albeit expensive way to manage the process and make sure that the system is being built to the owner's expectations. The disadvantages of this approach include loss of staff from day-to-day activities while at the integrator's facility and the possibility of staff assigned to the facility having a different idea or goal than the owner.

System integrators frequently will work at an owner site to configure the control system processor and graphics. Working in close proximity is one way to involve the owner in all steps of the process and to get input and quick feedback for some of the more subjective phases of controls implementation, such as graphic development. With the owner onsite to review and provide quick feedback, the use of color, abbreviations, phrases, shapes and symbols, and screen navigation required for graphic development can be defined quickly and efficiently.

One facility (Massachusetts Water Resources Authority), which had never had any type of automated controls, used the factory demonstration test as an introduction to the operators. Several key operators attended the shop test and participated in the system test. The hands-on environment gave them a first look at control graphics and the control keyboard and a chance to become familiar with both the hardware and the graphics. The positive experience enabled the operators to return to the plant with an enthusiastic attitude about the new control system. They were able to demystify the system for other operators and really helped to develop a positive attitude.

4.10.2.3 Installation

Many of the owners described establishing a successful long-term relationship with installing contractors who had extensive experience with I&C installation. The benefit of working with the same contractor was that the installations would be consistent and the owner would have a degree of confidence that the approach and workmanship would meet expectations.

In some cases, the owner would take over the inspection process or take a significant role in the process.

4.10.2.4 Field Testing

Successful systems require complete testing of the installed system using a systematic approach. Several of the owners described testing their own systems; others adopted an oversight role and watched work done by contractors and an engineer or construction manager. The projects with the most comprehensive test plans generally had the best overall results. Testing needed to include wiring testing to verify terminations, loop testing to confirm the integrity of control loops, and then control strategy testing to verify correct operation of configuration, and finally system testing, where the controls operated the process.

4.10.2.5 Startup

Startup needs to wait until testing is complete. One of the worst turnover situations occurs when an incompletely tested system is turned over to operations for use. The operations staff never gains confidence in the system after having a bad operations experience and never operates the system in automatic.

To prevent this scenario from occurring, complete testing of the system and process should be performed prior to turnover for full operations. MWRA, for example, integrated the process control system into the overall process test plan, so that water testing of the process was completed with the process control system operating the process in automatic (Laquidara et al., 1998).

4.10.3 System Maintenance

System maintenance includes several different types of tasks, including hardware maintenance, software maintenance, and system configuration changes. No owners described doing a cost-benefit or ROI study on the steps required to maintain systems. Priorities for completing projects were not specifically defined or justified through cost benefit analysis.

Very few utilities described using their Management Information Systems (MIS) departments to support control equipment and networking systems. Although new types of systems use similar technology, operating systems, and communications networks as enterprise systems, the process control staff is not relying on the MIS departments as a resource to maintain equipment.

4.10.3.1 Hardware and Instrument Maintenance

System hardware maintenance responsibilities varied: in some facilities the owner was fully responsible for system maintenance; in others system maintenance was contracted out. In several large plants, the owner had entered into 3- or 5-year agreements with the system vendor to maintain hardware. Maintenance included preventive tasks like cleaning the system, replacing expendable items like batteries and corrosion control modules, and corrective tasks like replacing failed modules and power supplies. The facilities that reported contracting out hardware maintenance reported doing so because they did not have sufficient staff to complete the activities in the vendor-specified amount of time. Other reasons for contracting out the work related to insufficient staff training. At facilities with several different types of control systems, the owner was not always trained to maintain all the systems.

The facilities with a wide range of systems installed described a high level of frustration with maintaining and servicing the systems. They described spending a significant amount of money on training staff to maintain systems, and then losing the staff and therefore the knowledge.

One major utility described contracting instrument calibration and maintenance for multiple facilities. The contract provided the owner with some assurance that instruments would be maintained and available for process control. The contract spanned several years, with the staff adding one or two plants to contract maintenance each year. When the contract became too large, the utility refused to renew the contract. The end result has been an increased number of control loops operating in manual, due to failed or uncalibrated instruments.

4.10.3.2 Software Maintenance

System software maintenance includes installation of patches and software upgrades. The software maintenance was completed either by service agreement or by staff, depending on contractual/warranty agreements, maintenance agreements, and owner staff training and familiarity with the system.

Many software upgrades at the field level required board level component changes. Owners would typically forgo making the change, or contract with the system vendor for them to complete the work.

4.10.3.3 Software Configuration

System configuration or reconfiguration was performed through a variety of mechanisms. Owners frequently have staff familiar with the steps required to configure or change system process control configuration. Owner staff would make changes and optimize control strategies originally developed as part of a contract.

Significant system changes, including process changes and major reconfiguration work, however, was frequently contracted out to system houses or to system integrators, because staff had too many other responsibilities to take on the added workload of major changes.

4.11 Operator Interfaces

Some owners felt that they had locked themselves into a specific system when they selected a single technology available several years ago, and are now unable to upgrade or expand the system to take advantage of new technology and latest operating and networking systems. In all cases the systems were still working. The owners were reviewing most current technology, and wishing for a simple way to port their existing system to the new technology.

During the 1980s and early 1990s workstation software was written for Unix or VMS operating systems. In the past few years many new operator workstations are written for Windows NT, which offers standardized maintenance and setup, and simplified networking. The owners with Unix- or VMS-based systems are using the NT-based software for new installations but are finding that all operator screens need to be redeveloped from scratch – a job much more significant than at first considered, since it requires retesting of every point to the graphics.

4.11.1 Workstation Technology

Systems include a wide range of technology. Many systems still operate with workstations running character-based graphics that typically have low-resolution graphics cumbersome to change, with very limited graphic depiction capabilities. Current systems use pixel graphics and allow importing images from a variety of formats, providing much more flexibility in graphic development.

Many of the systems with character-based graphics were implemented in the 1980s and are currently scheduled for replacement.

4.11.2 Integration with Other Information Systems

While many utilities reported plans to integrate their process control data with other enterprise systems, few utilities reported having completed the integration of process control data with other systems available for the utility. The other systems include asset management systems, computerized maintenance systems, and operational data reporting systems. The most advanced integration was with the operational data reporting systems.

4.11.3 System Selection

Operator workstation selection typically was completed as part of the process control system procurement. Owners described a process involving an evaluated bid, where technology was reviewed and then selected based on the entire system. Once a system was selected, it became the system selected for all facilities. The exception to this was at one large utility with multiple treatment facilities. To take advantage of new technology, this utility changed technology at the operator workstation level for newly automated facilities. The change precluded integration with existing SCADA, but the utility was willing to forgo the integration for the advantages of the newly available systems.

4.11.4 System Implementation

Based on the interviews, the implementation of operator workstation graphics is one of the most reworked phases of implementation. Many owners described revising or completely redoing operator graphics once they'd taken ownership of the systems. Owners generally described a desire to reduce the level of effort and rework with each system.

Strategies for reducing rework included contracting with the same system integrator for multiple facilities. By getting the same staff involved for multiple projects, repeat staff would use similar implementation techniques and reuse code and graphics. Another utility required that the system integrator start with code and graphics developed for operational facilities. Many utilities described using standards to guide system integrators with their work.

One utility's strategy was to budget for the rework. It increased the budget to include 6-9 months staff time to modify and redo operator workstation graphics.

One area most frequently neglected in the implementation of operator workstation configuration was alarm handling and event logging. Although most operator workstations reported robust alarm and event packages, many utilities had not implemented the systems to benefit their operations staff.

Chapter 5.0

KEYS TO SUCCESS

5.1 Introduction

The planning, design, implementation, and operation of a control system is a continuous process, one that takes place over many years. It is also an evolutionary process, constantly changing as owners and users become increasingly educated and computer savvy, process and organizational needs change, and technology advances. All of these dynamics challenge control system projects to remain successful throughout their planned lifetime.

During the WEFTEC 2001 workshop held for this project, attendees were asked the following questions and provided the following responses:

Question	Not enough	Enough	Too much
Are we doing enough with treatment plant automation?	70%	30%	0%

Question	Not even close	On the mark	Far exceed
In your opinion has automation lived up to its potential?	54%	46%	0%

These responses, from a group of leaders providing automation at wastewater treatment facilities, identify an opportunity and a need for improving process control systems. The following discussion presents practices both to be taken and avoided in order to avoid the pitfalls that frequent control system projects.

During one of the group exercises at the 2000 WEFTEC meeting, attendees identified factors they considered contributory to control system success. A tally of the success factors identified by the workshop attendees and the facilities visited during the field surveys resulted in the top 10 success factors shown in Table 5-1.[1] The list starts with the most frequently occurring success factor to the 10th most frequently occurring factor. The list was validated during the WEFTEC 2001 workshop where 39 attendees (93%) agreed with the list, while 3 (7%) disagreed. The following table lists the success factors and phase of the project where they are especially important:

Table 5-1. Key Success Factors and Their Importance During Each Phase of a Project

Success Factor	Planning	Design	Implementation	Operations
System Advocate	X	X	X	X
Management Support	X	X	X	X
Owner/Operator Involvement	X	X	X	X
Trust in Operators	X	X	X	X
What's In It For Me	X	X	X	X
Plan for Maintenance	X	X	X	X
Full System Testing			X	X
Preparation for Failure	X	X	X	X
Training			X	X
Setting Goals	X			X

Almost all the success factors on this list are organizational and management rather than technical issues.

The following sections describe these "keys to success," which should be implemented throughout the lifetime of the system. The list is not complete, nor does it guarantee system success when followed, but it does indicate some of the common success factors.

5.2 System Advocate

The success or failure of any automation project depends on the people leading it. In the past, successful projects have had a system advocate, someone who was technically competent, really excited about the project, and respected throughout the organization. Since staff turnover rate is high in organizations, the role of system advocate should be assumed by several people who work as a project task force. The project task force should be the team responsible for taking the project from planning through operations. The team provides continuity should one member decide to leave the organization. By working together from the start of the project, the team develops a common and consensual understanding of the system requirements, capabilities, and features.

The task force should be led by a key member from management and should include representatives from operations and maintenance groups, engineering staff, and information technology (IT), though the team should not be led by IT staff. The team should include a project manager who is responsible for logistical issues related to managing the team.

[1] The process of tallying the success factors with the facilities involved reviewing each facility and relying on the impression of the WERF Team member, based on the onsite field survey.

5.3 Management Support

Successful projects must have the full backing of management. A high level statement committing the organization to the project, and the project task force, is an essential success factor.

Management needs to make a financial commitment to the project, both for the cost of procuring and maintaining the system and for backing the project task force.

Management must commit to the organizational change by putting resources into the project task force and by supporting the integration of the system into the organization. If the system will result in reassignments or reduction in staff, then management needs to develop and communicate the plan for reorganization. For example, in the case of staff reduction, management needs to state the method for reducing staff, such as only through attrition.

5.4 Owner/Operator Involvement

Large control system projects are usually planned, designed, and implemented by consulting engineers hired by the owner/operator. The level of effort required to complete the project, combined with the risk, make control system projects well suited for contract work. The disadvantage of having the work completed under contract is that the owner/operator's needs and intentions are not always met by the project.

The phases of the project where owner/operator involvement are particularly important include initial project definition. The project objectives must be based on the owner's needs and expectations. Once involved, it is the owner's responsibility to stay involved through every phase, and to validate that the system is being built to meet the needs in the stated objectives.

During the design phase, owner/operator involvement is critical to establishing the overall goals for the controls (Stire, 1983). The engineer is responsible for the details and accuracy of the design, but the owner should review design documents for O&M-related issues.

During the implementation phase of the project, owners and especially operators should be involved in system testing, both in the factory and in the field.

5.5 Trust in Operators

The ultimate indication of a system's success is whether or not it is used for its intended purpose. Creating a system that meets operators' needs is the best way to get them to use the system. By involving operations staff in all phases of the project, and trusting them to state what they need, the likelihood of operator buy-in to the system increases significantly.

During the planning phase of the project, operations involvement in the project task force is important to obtain input and feedback related to the overall operability of the control system. Operations staff will provide critical input related specifically to issues like ease of use of the control system and functionality. Operations provide a perspective frequently overlooked or understated by engineering and management staff. Operations require simplicity in the system that other members of the project task force may not appreciate.

During the design phase, operations provide valuable information needed to define control strategies. For a plant retrofit, this is especially important. It is also important to verify and validate information received from operations staff.[2]

During the implementation phase of a project, operations staff must be integrally involved in testing the system. Involvement during testing provides a critical opportunity to introduce operations staff to the system and give them a chance to use the system without adversely affecting the treatment process.

Finally giving operations a chance to provide input regarding the operating system provides one means for improving the system. Ongoing use of the system depends on Operations staff being involved in system improvements.

5.6 What's In It for Me?

Every party having a stake in a process control system will want the system for different reasons. Meeting the needs of the different parties increases the likelihood that the system will be accepted and considered a success. It is also important to pick stakeholders who have a broad enough vision that their needs are representative of the goals of the entire organization.

Of course, the parties that have a stake in the system need to be part of the project task force responsible for the system from the planning stage on. The project objectives need to reflect the specific needs of each of the stakeholders. Ongoing stakeholder involvement, through each stage of the project, is one way to make sure that the system is implemented to address each of the objectives.

Addressing "What's in it for me?" for each stakeholder is the best way to keep the stakeholder involved and committed to the success of the project.

5.7 Plan for Maintenance

How reliable does the system need to be, and what steps need to be taken to maintain that level of reliability? System availability is a tangible measure of reliability: how long can the process run without the control system. Availability is calculated as a function of the reliability parameters (mean time between failure and the time to repair) of the individual components of the system. But in order to achieve the planned level of reliability, equipment must be maintained based on the manufacturer requirement.

During the design phase issues related to system reliability need to be addressed, including redundancy, power source, power backup, surge and lightning protection, control response to equipment failure, and location and accessibility of equipment to name a few. During the design phase aspects of a reliable system need to be incorporated into the design

---

[2] Plant retrofits may require revising instrumentation and controls to change the way operators run a process. In such situations, either because operations have been running it incorrectly, or because there's a process change, during retrofit, a major change in the control strategy may be needed to change the method of operation.

document, and a method of monitoring the failures needs to be addressed. System monitoring packages are generally available, but need to be provided as part of the design package. Provisions for setting up and configuring the package should also be addressed in the design.

During the construction phase of a project, inspection installation should include review of equipment for ease of access, maintenance, and replacement. The construction phase also needs to include provisions for maintaining equipment until turnover to prevent voiding of any warranty. Once equipment is turned over, ongoing maintenance can be by plant staff or by system suppler under contract. Although expensive, contracting system maintenance is one way to make sure the steps are completed.

5.8 Full System Testing

Complete system testing and sign-off is a critical step to the perceived success of the control system. When a system is first turned over for operation, any aspect that doesn't work is viewed as a failure and contributes to an overall negative perception of the system. Planning for full system testing must be addressed during every phase of the project, so that the necessary tools and documents are available to support the testing process. The full system testing needs to be witnessed by owner representatives, because unwitnessed testing is frequently compromised when budget or schedule get tight for other reasons.

During the planning phase, full system testing includes allocating sufficient budget, resources, and time so that testing can be completed prior to completion of the project. The resources, including owner staff, need to be authorized, through the planning stage, to participate in all testing activities.

During the design phase, the requirement for contractor testing needs to be incorporated into contract language so that the contractor's role and responsibility is clearly defined. The design document also needs to establish the requirements for the system, including control strategies.

During implementation, testing needs to include a complete factory test of hardware and software, field-testing of all process I/O points to the end element or instrument, and complete system testing for operation with the process. Testing plans need to have flexibility, so that testing will have the least overall impact on project schedule. For example, one project called for a full system factory test, including both hardware and software. Delays in the submittal process for control strategies resulted in hardware being ready, but control strategy configuration not being complete. In this particular case the requirement for factory testing was modified to allow two tests. The first test was for the hardware only. Once complete the hardware could ship and be installed. The software followed and was tested in a separate system. The split test allowed onsite installation and wiring to proceed while providing additional time to finalize and test the control configuration.

During operation, changes to the system need to be fully tested before turnover to operations.

5.9 Preparation for Failure

Despite the best intentions, not every aspect of a project goes exactly as planned. Contingency planning, and creative solutions are essential to getting derailed projects back on track. Flexibility in project budget and schedule provide the best means to absorb the impact of an unwanted, unexpected issue. Flexibility in the project task force approach is essential to develop creative workarounds to any unplanned issues.

5.10 Training

The best of system implementations will not be of any benefit if the user is not trained to use the system as intended. The training budget needs to include both the cost of training and allowance for staff to attend training. The overall training program needs to provide varied methods to maximize training by allowing for different student learning styles. Many different training options exist, including train-the-trainer, formal classroom training by the vendor, hands-on, self-directed, and on-the-job.

Very large system implementations, especially ones that include a major process upgrade, may require a separate training shift to ensure staff training is complete. With a training shift, staff is scheduled to attend training during daytime hours and is relieved from regular responsibilities. Scheduling a training shift does require increased payroll and workload, but it ensures staff will be available for and will attend training. Small system changes will easily be addressed by training in the control room during normal operating shifts.

Training should be scheduled prior to system turnover with follow-up refresher training after the system has been operational for several weeks. Some type of repeat or follow-up training should be available at regular intervals thereafter to provide refresher courses, and for new employees.

5.11 Setting Goals

Before starting any project, a written project plan needs to be developed. The plan forms the basis for the entire project and should be referenced for the duration of the project, whenever there are questions of intent, goals, schedule, and priorities. The written plan, authorized by management and key personnel, represents a consensus for the control system. Different consultants use different planning methodologies, but the methodology used to capture the plan is not important so long as the plan defines the objectives. A statement of objectives that defines both tangible and intangible goals becomes a justification for the project and a measure to determine the success of the project. The goals need to define the specific benefits to be derived from implementation of the system.

5.12 Barriers to Success

Rigid procurement methodologies sometimes restrict a project team's ability to select a control system and control system supplier on any criteria except cost. This limitation can force selection of a system that is marginally suited for the project requirements, compromising the ability of the system to succeed. The procurement approach should be addressed and resolved prior to the development of design documents. An evaluated technical bid provides the most flexibility when procuring a new system.

Incorrectly sized equipment, including flow elements and valves, occurs frequently when non-instrument engineers design instrumentation. Incorrectly sized instrumentation cannot be controlled with any degree of accuracy and is the result of many improperly operating control loops.

Using instrumentation in control system projects requires both selection of the appropriate sensor and installation based on manufacturer guidelines. Faulty or unreliable measurement is more destructive than no measurement at all, because it sets a bad precedent and destroys the confidence of the operators department. Sufficient care should be taken to either remove or repair any faulty instrumentation.

Chapter 6.0

TRENDS

6.1 Introduction

The instrumentation and control survey of 110 treatment facilities and the field surveys of 25 industrial and municipal treatment works were intended to document the current state-of-the-art in WWTP sensing and control systems. At the same time, discussions with the management and operations staff at these facilities, along with reviews of current technical literature and workshops conducted at WEFTEC 2000 and WEFTEC 2001, provided the project team with valuable insight into the direction that WWTP sensing and control systems are headed. Future trends in the areas of sensor technology, process control strategies, control system technology is discussed below.

6.2 Sensor Technology

A WERF *Research Needs Assessment Study* conducted by ITA (WERF, 1994) concluded that on-line sensors were commercially available to measure most contaminants (organics, metals, other inorganics) in an aqueous environment. For the few contaminants for which prototype or commercial instrumentation was not available, research activities were generally ongoing to develop such sensors. The study also showed that virtually any analytical methodology available in the laboratory could be or had been modified to be used in an on-line application. A major constraint to commercialization of these sensors is the lack of a market of sufficient size to warrant the necessary investment.

The instrumentation and control survey done for this report found very few applications of sophisticated wet chemical sensors in WWTPs. The most sophisticated on-line sensors were those that measured ammonia, suspended solids, or BOD. This finding was confirmed by the field surveys where, even at the facilities using sophisticated control strategies, parameter specific sensors were not commonly used. WWTP operations and management staff often expressed concerns about the reliability, robustness, and life cycle costs of more sophisticated sensors. Despite these concerns, however, some interesting trends in sensor technology are emerging.

6.2.1 Sample Transport and Conditioning

The single major deterrent to the use of more sophisticated on-line, parameter specific sensors that depend on a wet chemical analytical methodology for measurement has been the lack of dependable, robust sample transport and conditioning equipment. Many manufacturers provide this equipment. However, its use in a WWTP environment has been less than exemplary as documented in a recent ITA test of on-line ammonia sensors (ITA, 2001). Sample pumps have a tendency to plug. Equipment provided to filter the sample prior to analysis has also proven to be problematic.

To address this problem, some manufacturers are now developing and offering in-situ sensors that do not depend on a sample transport and conditioning system. These sensor systems, if properly designed and installed, can eliminate the need to transport the sample to the sensor, reducing the risk of sample loss or a change in the analyte in the sample transport system. These sensor systems can also mitigate the problems with plugging of sample lines and sample filters. Such sensors are available for measurement of chlorine residual, ammonia, nitrate, phosphorus, and other parameters. More sensor manufacturers are focused on adapting their sensors for in-situ use because of the perceived advantages.

6.2.2 Analytical Sensor Technologies

The increased emphasis on nutrient removal (nitrogen and phosphorus) in WWTPs will increase the use of commercially available on-line sensors to measure parameters such as ammonia, nitrate, and phosphorus on-line; however, WWTP management and operating staff must be convinced that these sensors can provide reliable information at a reasonable cost. The case history of the Collingwood WWTP presented in Chapter 4 of this report is an indication that on-line measurement and automated control using this instrumentation is achievable. This instrumentation is much more commonly used in Europe than it is in North America. Performance testing by organizations such as ITA or by individual users is essential to demonstrate that this instrumentation can provide value in WWTP applications.

Using on-line sensors to measure such parameters as heavy metals or specific organics in wastewater treatment plants is very rare. Similarly, on-line sensors that measure biological activity at the cellular level (biosensors) may have applications in research but have not found widespread application at the operational level. Dependable sensors able to provide operational staff with reliable information that would benefit decision making will need to be proven in real-world applications before wider application of such technologies will occur in the wastewater treatment industry.

6.2.3 Mechanical Equipment Condition Sensors

Trends toward satellite operation of remote facilities and darkened control rooms in WWTPs, as discussed elsewhere in this chapter, will depend on the ability to continuously monitor and report on the operating condition of mechanical equipment. The process industry has made extensive use of sensors for vibration analysis, oil analysis, thermography, and motor current signatures. The output signals from these sensors can be utilized in the industries' preventative maintenance and CMMS programs (Davison, 2000). These technologies have been little applied in the wastewater industry, but will be essential to replace the manual "hear and touch" approach that has been a fundamental part of WWTP operators' normal routine.

6.3 Process Control Strategies

6.3.1 Control of Satellite WWTPs

The trend toward satellite operation of remote facilities and darkened control rooms in WWTPs will require a more sophisticated level of monitoring and control to ensure that compliance can be maintained in these unmanned facilities. Currently, these operations generally depend on simple monitoring of the operating condition of key equipment such as pumps and blowers, and alarming to inform remote staff of mechanical or electrical problems. As the trend toward unmanned operations increases due to economic and competitive pressures, more sophisticated process monitoring will be needed to inform the remote operating staff of the operating condition of the facility. This will include sensors to measure the concentrations of specific parameters such as suspended solids, ammonia, phosphorus, and chlorine that are regulated under the facility's discharge permit. Increased use of these sensors will allow more sophisticated process control strategies to be implemented as an initial response to a measured change in performance.

6.3.2 Simulation Models for Process Control

Simulation models based on the IAWPRC activated sludge model and incorporating sophisticated dynamic models of secondary clarification and other WWTP unit processes have become an essential tool for process design of WWTPs and are becoming common in operator training applications. These simulation models can accurately predict the effect of changing an operating condition on the performance of a plant.

Research at the full-scale plant level is now being undertaken to link these process simulators to the plant SCADA system and use the models for process control. As changes in operating conditions in the plant are measured with on-line sensors (flow, DO, MLSS. SRT, etc.), the dynamic models can predict the impact on performance and provide output signals through the SCADA system to change key operating parameters and maintain the desired performance. Integration of the process models into expert systems that can learn from their experience and effectively troubleshoot WWTPs will be the next step in this development.

6.4 Control System Technology

Rapid advances in the technology available for computing has had a profound impact on all types of business over the past two decades. The development of more powerful computers in ever-smaller packages has allowed computing and technology industries to grow in ways never before imagined. The controls industry has benefited significantly, evidenced by the articles, advertising, and product promotions on the pages of virtually every control system trade publication.

The municipal wastewater market has traditionally been a conservative market compared to other industrial markets, reluctant to leap to new, untried technology, and preferring to adopt a wait-and-see attitude. This conservatism has allowed the industry to avoid investing in some expensive mistakes; it has also, however, delayed the industry's foray into some of the newer technologies.

Industry standards are now driving much technological advancement, making the investment and transition a "safer bet" than some of the older technologies. The following are several areas where changes in technology have or will provide significant benefit to system owners:

6.4.1 Ethernet to the Instrument

One of the greatest areas of growth relates to the interface to field instruments. As bandwidth has increased and security advanced, Ethernet is used more and more to link control equipment.

Some field instruments are advertised as Ethernet-enabled; however, it is not always practical to extend Ethernet cabling all the way to the end element. Ethernet to an I/O rack or remote I/O is a common implementation. Field instruments are wired to the I/O rack using conventional 4-20 signals or Fieldbus, Profibus, or one of the other instrument network protocols.

6.4.2 Wireless Technology for Communication for Instrument Hookup and Operations Interface

Wireless technology provides advantages both at the field end and at the operator end of operations. It only a question of time before the technology really takes off.

Spread Spectrum radio has already been used frequently in process control applications. Spread Spectrum radio allows relatively low cost, unlicensed communications between facilities. Although it is typically used for communication between geographically separated facilities, it can be used intra-facility where lack of conduit/duct bank capacity precludes additional electrical connections. Spread spectrum is implemented with point-to-point communications.

Emerging technologies like Blue Tooth and 802.11b, both wireless systems using the 2.4 GHz bandwidth for connecting computers, printers, and other remote devices, are battling to determine which technology will become the leader. Neither has been incorporated in the process control market. However, they are both available for equipment used in the control room. If the standards take hold in the business world, then they may start to appear first in control rooms, since the distance of operation is limited by restrictions on the power output. As the leader emerges and one of the technologies takes off, this power output restriction will probably be overcome.

New products available in the market include Ethernet equipment equipped with remote-mounted RF transmitters and receivers. As this type of product becomes available, more and more instruments will be equipped with RF communications and flash memory, allowing the user to configure the network and flash the communications protocol into the device.

6.4.3 Data Sharing to Other Enterprise Applications

End users are demanding integrated data solutions such that process control data, computerized maintenance management systems, financial packages, and laboratory data are all available throughout the organization. Systems vendors continue to promise that their systems are open and compatible and can be integrated with a wide range of other systems.

The reality of integrating the data extends far beyond the technical challenge of linking databases. Most important is the technical challenge of setting up the systems so that the data from each different application is available throughout the network in such a way that it is easy to locate and use.

6.4.4 Data Sharing with Customers

Data sharing between businesses has become a frequent occurrence, as a part of business-to-business transactions. The water and wastewater industries have not yet started widespread sharing of data; however, the tools are available from industry. Applications that allow data sharing across the Internet make it easier to exchange data between organizations.

6.4.5 Common PLC Programming Standards

Programming standards have been used to varying degrees since the start of microprocessor based controls. The discipline required to develop and maintain standards and then refer back to them for every project provides significant benefits by reducing the time required to develop configurations and to simplify the PLC program modification and code maintenance.

The extent of standardization can vary from general guidelines for coding and naming all the way to reusable code segments. Vendors and system integrators use standards and code segments to make the task of programming controls more efficient. Most utilities have some form of programming standards in place. As programming tools become more sophisticated, they provide improved methods for cataloging and reusing code.

Organized standards, like IEC 1131, provide some general framework for PLC programming, but no common PLC control logic exists for all PLCs. In the future, third party software developers have the opportunity to create high level PLC programming language that can be compiled to run in any type of PLC.

6.4.6 Proprietary Application-Specific Control Systems

Application-specific controls are available to optimize equipment and processes. The vendor packages are frequently required to protect the intellectual property associated with the equipment or the process. Vendor-supplied control packages also frequently require vendor involvement to reprogram or tune the system. Simply increasing the process load may require retuning of the controls.

6.4.7 Satellite Operation of Remote Facilities

More and more, utilities are operating remote facilities from a remote site. Judicious use of remote monitoring and control allows facilities to be operated in a reliable way, while significantly decreasing staff costs. The use of Web cameras (video transmitting images over the Internet) has not been deployed extensively in this country yet for the purposes of monitoring wastewater facilities. However, as the use of video increases, it will provide significant benefit to water and wastewater utilities by allowing them to skip site visits and save on travel costs.

In order for the remote operation to be successful, the controls must be reliable. One utility has developed a reliable enough control system that it visited the remote sites only once every other week.

6.4.8 PLC Networks Compete with DCS Systems in Functionality

The significant increase in computing power over the past few years has increased the functionality and versatility of PLCs to the point that they are now nearly as powerful as distributed control systems (DCS), while providing many additional benefits, such as compatibility with more open systems, like Ethernet. DCS systems are also opening their architecture to allow adding PLCs into the system. The change in technology is allowing owners greater flexibility when selecting systems for control system expansion and replacement.

In the past, DCS has offered much more powerful configuration tools and capabilities than PLC programs. Users are now demanding greater sophistication in PLC logic, forcing vendors to make the software capabilities of PLCs as powerful as DCS.

6.4.9 Internet/Intranet Tie-In

Browser-based operator interface has greatly extended flexibility and access to process control systems. The browser-based systems can use existing facility local and wide area networks to put the real-time data from the process control system on any employee's desk. With appropriate security measures in place, the same information can also be made available over the Internet.

6.4.10Manufacturer and Owner Monitoring of Remote Equipment

A recent addition to the service market involves instruments equipped with modems. Manufacturers are able to monitor the status of the equipment from a remote site and determine when service/maintenance is required. The manufacturers monitor the instrument via phone line, or wireless connection, and contract to provide all maintenance services. The cost and benefit of this type of arrangement has yet to be analyzed.

Manufacturers are also providing remote data collection over phone or Internet connections. Tank monitoring instruments, for example, monitor on-hand inventory as a means to determine when delivery will be required.

6.4.11Training – Distance Learning

Technology has changed the methods available for training by adding many new options. Distance learning, using on-line conferencing, video, audio, and the Internet, makes training available on demand.

6.4.12 Centralized Laboratory or Contract Laboratory Data

Virtually all laboratories, whether private contract labs or publicly owned and operated labs, that provide analytical services related to the monitoring of WWTPs are equipped with laboratory information systems (LIS) that can report data electronically to the client. This data can be inputted directly into the WWTP's data management system to facilitate report preparation and to be accessible to all management and operations personnel.

6.4.13 Simplified Computer Interface for Both Input and Output

Process control systems now use standard computers and software. The advantage to control system vendors is that it is easy for them to add operator interface devices as they become available on the open market. In the past proprietary systems had limited interfaces, using devices like a mouse or trackball. Windows-based systems have a wide range of operator interface devices, including voice input and eye motion tracking devices.

The open architecture will allow control system computer interface to stay current with new operator interface products.

6.4.14 Three-Dimensional Graphics

Computer graphics have improved significantly since the character-based graphics available with the first control systems. Three-dimensional graphics are now used frequently in plant design documentation. Three-dimensional graphics can be used to present operator graphics in a way that increases the operators' comprehension of the process and process equipment as well as the process data. The next step will include integrating control system graphics with other applications, including O&M Manuals, GIS, SOP, and CMMS.

6.4.15 Darkened Control Rooms

As controls and communication become more reliable, the reliance on operations staff to watch what is going on in a plant will decrease. Alternative methods of monitoring the process, through pagers and handheld units will allow operations staff to spend time in the field, maintaining equipment

6.4.16 Handheld Wireless Interfaces

Business applications using wireless connections to cellular phones and personal data assistance (PDA) make real-time and near real-time data available to subscribers. The logical extension of this technology to process control includes transmission of process alarm information to remote users. For example, if the plant effluent chlorine residual measurement drops below a set value, an alarm would be generated on alphanumeric pagers worn by the plant manager and the chief operator.

The limited size of screens on cellular phone and PDAs does reduce the amount of information that can be transmitted and easily viewed. Another barrier to successful implementation relates to the technical "savviness" of the user on the receiving end of the alarm or message.

In order for this type of interface to be really effective, the information provided needs to be specific information needed by the user, and it needs to be presented in a manner that makes it easy for the user to determine what the issue is and what actions are necessary.

Chapter 7.0

FUTURE NEEDS

7.1 Introduction

As noted in Chapter 5, the majority of participants (21 out of 39) at the WEFTEC Workshop held in 2001 indicated that automation had not lived up to its potential in WWTP applications. Discussions at this workshop and with WWTP management and operating staff at the utilities and industries that participated in the field surveys identified some specific needs that must be met if the full potential of WWTP sensing and control systems are to be realized. These needs are discussed briefly below.

7.2 Sample Transport and Conditioning Systems

On-line sensors are currently available that accurately measure the concentrations of the critical WWTP process control parameters. These sensors perform well in a relatively clean environment and require relatively little ongoing maintenance. However, a WWTP is a severe environment, and the failure of many of these sensor systems to perform as advertised is often related to an inability to effectively transport and condition the sample. Recent testing by the Instrumentation Testing Association on ammonia analyzers (ITA, 2001) and discussions with WWTP operating staff confirmed that this remains a major problem.

Considerable effort has been expended in the development of sensor technologies. Comparable effort is needed to develop effective and reliable equipment to transport and condition the sample for analysis before WWTP operating and management staff will be willing to depend on the available sensor technology.

7.3 Instrument Testing

Much of the available information on the performance of available sensors is anecdotal. During the field survey, when WWTP operating staff was asked why they were not making full use of the sensors installed in their facilities, the most common answer related to the unreliable nature of the sensors. This problem has plagued the industry since the first instrumentation and automation surveys in the 1970s. As a result, the perception of poor instrument performance is perpetuated and the lack of confidence continues. No consideration is given to whether the proper instrument was selected for the application, whether the installation was appropriately designed, or whether the sensor was properly maintained.

The only way to change the perception that the available instrumentation is not adequate is to conduct rigorous testing to detailed protocols and procedures and with appropriate quality assurance and quality control. This testing must involve both the manufacturers and the users, and the results must be effectively disseminated to the industry. This approach will ensure that sensors that do not perform are identified so that manufacturers can make necessary improvements. More importantly, sensors that are reliable can be identified so that users can have some confidence in their performance at the time of purchase.

Many utilities currently undertake their own expensive analyzer testing. Through a coordinated effort, this testing could be improved to ensure that the results are valid and can be replicated. An information clearinghouse is necessary to ensure that all valid testing information is available to the users.

7.4 New Sensor Technologies

This investigation demonstrated that the major focus needs to be placed on ensuring that the available sensors are reliable and robust, rather than on the development of yet more exotic sensors to measure other parameters. However, a few specific sensor technologies warrant further research and development

The bane of WWTP management staff everywhere is odor. The ability to dependably measure odor on-line continuously would not only provide early warning of odor incidences, but it could be used to detect odor sources and potentially to activate odor control efforts. Instruments to measure some odor-causing constituents such as hydrogen sulfide are available commercially, but more research is needed on sensor array technology or "electronic noses" before this technology is ready for use. The ultimate goal would be automation of odor control systems based on such on-line sensing equipment.

Successfully unmanning and remotely monitoring satellite WWTPs requires improvements in sensor technologies to monitor the physical status of mechanical equipment. Building these sensors into the original equipment would improve reliability and reduce cost.

Biosensors and related sensor technologies able to provide early warning of upset conditions in the biological process have been developed at the research level but have not proven dependable for continuous use. Further work is needed to develop robust, reliable sensors able to operate cost effectively in a full-scale wastewater treatment plant environment.

7.5 Demonstrated Benefits of Automated Control

Despite the perception that automation of WWTPs has not achieved its potential, the field surveys identified many instances where automation had realized significant benefit. Many users justified automation by estimating the economic benefits and making a business case. However, few utilities have effectively followed up by monitoring the cost savings and demonstrating that the original assumptions were valid. A thorough comparison of the costs and performance capability of a WWTP before and after the implementation of a comprehensive automation system is needed to convince the industry of the potential for WWTP sensing and control systems.

7.6 Better Education for Designers, Suppliers, and Owners

Ultimately, the success of any automation project depends on the quality of the design, installation, and equipment selection, and on the commitment of the owner to maintain and continuously upgrade it. Failure at any stage will virtually guarantee the failure of the project.

Operating organizations should consider adding one or more certified instrument technicians to their operating staff, and organizations with substantial plants should consider adding a licensed control systems engineer.

Consulting engineers, contractors, equipment suppliers, and owners all need to better understand the role that each plays in the successful implementation of a WWTP automation project, their responsibilities at each stage of the project, and the benefits that can be realized by successful WWTP automation.

APPENDIX A

Instrumentation and Control Survey

The following survey was sent to all Water Environment Research Foundation subscribers and all Instrumentation Testing Association members to assess the current state-of-the-art of automation at a wide variety of wastewater treatment facilities. The survey was advertised and available on ITA's web site and was also sent to a small number of other facilities known to the research team. The survey was used to assess the current state-of-the-art of automation and control at existing plants and to help plan the field survey portion of the project.

Out of approximately 300 surveys mailed, a total of 110 surveys representing 45 utilities were returned. Each survey represented results from a single treatment plant. Several utilities responded for multiple facilities. The City of Houston, in fact, responded with 41 surveys. Kansas City responded with seven. The large number of surveys from Houston likely skewed some results, especially in the area of partially attended operation of plants. Any known or suspected skewing is discussed in the survey results.

Complete survey results are available in an Access database and a series of Excel spreadsheets.

The survey results are discussed in Chapter 2 of this report.

APPENDIX B

FIELD SURVEY

The project team used the following survey to collect information on the current level of automation at 25 municipal and industrial wastewater treatment facilities. The survey was used to assess both successful and problematic automation practices and to document types of control strategies in use. The plants surveyed were:

- Anonymous paper plant – 3.5 mgd
- Blue Ridge Paper, Canton, NC – 60 mgd design, currently operating at 25 mgd
- Boise Cascade, International Falls, MN – 25 mgd
- Deer Island WWTP, Boston, MA, MWRA – 480 mgd
- Chicago Stickney WWTP – 800 mgd
- City of Las Vegas, NV – 66 mgd
- Clark County Sanitation District, NV – 88 mgd
- Collingwood WPCP, Ontario – 6.5 mgd
- Duffin Creek WPCP, Ontario – 96 mgd
- DuPont Cooper River WWTP, Charleston, SC – 1 mgd
- DuPont Dow Elastomers – Neoprene, Louisville, KY – 1 mgd
- Eastman Chemical Co., Kingsport, TN – 29 mgd
- Henderson, NV – 16 mgd
- Houston, TX - 69th Street WWTP – 75 mgd
- Houston, TX - Sims South WWTP – 36 mgd
- Houston, TX - Southwest WWTP – 25 mgd
- Louisville, KY - Cedar Creek WWTP – 2.5 mgd

- Louisville, KY - Morris Forman WWTP – 105 mgd
- Madison Metropolitan Sewerage District, WI – 40 mgd
- North Shore Regional Sanitation District, IL – 22 mgd
- Orange County NW WWTP, Orlando, FL – 8 mgd
- Roseville, CA - Dry Creek WWTP – 15 mgd
- San Francisco, CA - Southeast WTP – 80 mgd
- Santa Clara/San Jose, CA – 120 mgd
- United Defense L.P., Louisville, KY – 0.3 mgd (30,000 gpd)

The survey results are discussed in Chapters 3 and 4 of this report.

Plant Name:
Location:
Contact Name:
Phone:
Fax:
E-Mail:

1. Control System

(Prefill this information from ITA Survey and telephone interview prior to site visit).

Plant Size (ADF) ____________ MGD _____Peak MGD

❑ Combined Sewer ❑ Separated Sewer

______%Residential ________% Commercial _________%Industrial

I/O Count __________________Total

__________________ AI

__________________ AO

__________________ DI

__________________ DO

__________________ Calculated (virtual)

Staffing Information

- Organization Chart ❑ Copy Attached

  ❑ Union ❑ Non-union

- Number of:

  Admin/Management ____________
  Process Engineers ____________
  Instrument Techs ____________
  Electricians ____________
  Operators ____________
  Mechanics ____________

- Contractor Support (Vendor or other) ____________

▪ What part of PCS are you most proud of?

2. Control System

▪ Age of system/control system history (when was it built and/or modified, how long did it take to build and startup)

▪ Process narratives

  ❑ Available ❑ Attached

▪ P&IDS

  ❑ Available ❑ Attached

▪ Hierarchy/Backup/Redundancy (system configuration drawing)

  ❑ Available ❑ Attached

- Control hardware (name the system vendor/supplier and model)

- Where does control reside/Where do you operate from?

- Other documentation available (loop diagrams, panel wiring diagram, etc.)

- How much integration with other systems?

- What management information and reports are generated? Attach examples.

- Publications/Technical Papers

  ❑ Available ❑ Attached

- Attach photo documentation

3. Planning/Design/Installation Practices

- ☐ New system ☐ Retrofit/upgrade
- ☐ Stand-alone ☐ Overall Plant construction
- Was "business case" developed to justify?
  ☐ Yes (attach) ☐ No
- Describe Procurement Proc.
  ☐ Low Bid ☐ Evaluated Bid
- How involved was utility in the: (describe the role the utility staff performed)
  - Conceptual design
  - Final design
  - Equipment selection (instruments, final control elements)
  - Testing
  - Programming and installation
- Were standards used?
  ☐ Yes ☐ No
- Describe training provided

4. Commissioning/Operations Practices

- What commissioning problems occurred?

- How many hours of training are needed for new staff to independently operate the control system?

- Describe other specialized knowledge or training required to operate the control system

- What maintenance records are available

❑ CMMS ❑ Work Orders

❑ Others (please list)

____

____

____

- Who does maintenance?

❑ Contractor ❑ Programmers

❑ Electrician ❑ Process Engineer

❑ Inst. Techs ❑ Operators

❑ Laboratory ❑ Others

5. Cost and Benefits

- How much time does PCS maintenance take annually (in hours)?

- Quantify "down time" per year (down time constitutes time the process is not operational, or equipment must be operated "In Local")

- What is annual cost of parts/supplies/maintenance contracts for operating/maintaining PCS?

- What is estimated annual cost savings related to PCS?

  Staffing ____________________________

  Energy ____________________________

  Chemicals ____________________________

- Do you measure or report cost savings related to PCS?

  ☐ Yes ☐ No

  - How is this measured?

- Other benefits associated with PCS?

6. Process Area

- Identify Process Area Being Considered

  - ❑ Influent, Lift Station, and Collection Systems:

    - ❑ Manual Start / Stop or Adjustment
    - ❑ Varlable Speed Drive Pumps
    - ❑ Multiple Fixed Speed Pumps
    - ❑ Mixed Varlable / Fixed Speed Pumps
    - ❑ Automated Flow Equalization

  - ❑ Preliminary (Screenings, Grit Removal) and Primary:

    - ❑ Screening – Manual Start/Stop
    - ❑ Screening – Elapsed Time
    - ❑ Screening – Differential Level
    - ❑ Screening – Cumulative Flow
    - ❑ Grit Removal – Constant Flow
    - ❑ Grit Removal – Grit Density
    - ❑ Grit Removal – High Torque
    - ❑ Primary Sludge Pumps – Manual Control
    - ❑ Primary Sludge Pumps – Timers
    - ❑ Primary Sludge Pumps – Sludge Density
    - ❑ Primary Sludge Pumps – Sludge Depth

- ❑ Secondary:
  - ❑ Blower Control
    - ❑ Blower - Most Open Valve Strategy
    - ❑ Blower - Header Pressure Control
    - ❑ Other ______________________________
  - ❑ DO Control
    - ❑ DO - DO Only
    - ❑ DO - DO / Air Flow Compound Loops
    - ❑ Other ______________________________
  - ❑ Sludge Wasting Control
    - ❑ Wasting - Constant Flow Rate
    - ❑ Wasting - Constant MLSS
    - ❑ Wasting - Sludge Age / SRT / MCRT
    - ❑ Other ______________________________
  - ❑ Return Sludge Control
    - ❑ RAS - Constant Flow Rate
    - ❑ RAS - Proportional to Influent Flow Rate
    - ❑ Other ______________________________

- ❑ Step Feed Control
- ❑ Other ______________ Describe

❑ Solids Processing:

- ❑ Thickening – Method ______________________
- ❑ Stabilization – Method _____________________
- ❑ Dewatering – Method ______________________

❑ Effluent/Disinfection:

- ❑ Chlorination – Flow Proportional
- ❑ Chlorination – Residual Control
- ❑ Chlorination – ORP Control
- ❑ Chlorination – Compound Loop Control
- ❑ Dechlorination – Residual Control
- ❑ Dechlorination – ORP Control
- ❑ Dechlorination – Compound Loop Control
- ❑ UV – Flow Placing
- ❑ UV – Intensity Monitoring
- ❑ UV – Lamp Condition Monitoring
- ❑ Total Plant

7. Success Factors

- What components work well?
  - Why?
- What components do not work well?
  - Why?
- What would you do differently?
- What would you like to do that you can't do?
- What element are you most proud of?

APPENDIX C

INTERVIEW WITH F. GREG SHINSKEY

Greg Shinskey is a well-known industrial process control expert and author of several texts on process control. Mr. Shinskey also served on the advisory panel and granted the project team an interview to discuss the application of industrial process control experience to municipal wastewater treatment applications.

Excerpts from an Interview with Greg Shinskey

Where are more advanced instruments such as on-line GCs being used and how much maintenance effort is needed to keep these instruments performing well?

Shinskey: Since the WWTP is a process involving liquids and solids gas analysis, therefore GC is not required as part of the process (that we could think of). GC in industry is used for on-line analysis fed back to control systems. The best results come from analysis of low MW compounds, because they can be analyzed most quickly and the results used in the control system with little to no delay time.

GC maintenance is generally a function of the process stream being analyzed and is critical for accurate and repeatable operation of the instrument.

What types of sample preparation systems (filters, etc.) are being used in industry?

Shinskey: Sample preparation is required [this was stated with a smile]; it is generally a function of the stream being analyzed.

Are the semiconductor sensors that promise to give us everything (or at least many analyses) on a chip real?

Shinskey: Semiconductor sensors have been available for a long time. As long ago as the '70s people manufactured element-specific sensors and family sensors (the single element sensors were more accurate for a family than a single element sensor.) The only sensors that have survived (in that there has been a market for them) are those that found a specific application. In industry, people look for specific solutions to specific problems. Technology that works in one process has been developed specifically for that process.

What's in the future for instruments, data transmission, process control strategies, HMIs?

Shinskey: Information sharing is really important. In industry, each different type of plant has different requirements; sharing of information within the industry isn't necessarily going to benefit all, because everyone is faced with unique problems. Wastewater treatment, especially municipal systems, has very similar processes, and therefore should be able to share information to benefit all.

Process controls in industry: What types of control techniques do you see making the greatest improvements to the field of controls (advanced computational power in control systems for adaptive controls, fuzzy logic...)?

Shinskey: Fuzzy logic is not appropriate for process control for two reasons:

Conventional control systems are designed based on the real physical parameters that can be calculated for a process. Real calculations will control the process better than any fuzzy logic. Control systems using fuzzy logic may actually make a process unstable.

Fuzzy logic "dumbs down" controls so that engineers don't have to know the mass balances and energy balances associated with the process.

Statistical process control may have some application to WWTP. Based on the known and documented skewed distribution for the control output, the control setpoint could be calculated to allow the PV to exceed the quality limit, knowing that the product quality would still meet the specifications. [Shinskey developed an SPC algorithm based on a skewed distribution of control parameters, and the volume of the process downstream from the control location, and the quality limits for the process output.]

One potential WWTP application is disinfection and the NPDES permitting process. Typically, plants will control output to a percent (10%) of their permitted concentration, so that they will not violate permit. Permits are structured for compliance to be measured during a long time period, such as 30 days. SPC maybe be able to be applied to this problem, allowing plants to reduce their disinfection costs and possibly exceed permit for short periods of time but still meet permit requirements over the reported period.

What types of controls do you see misapplied most frequently?

Shinskey: Linear multivariable predictive control systems are oversold, designed, and maintained by consultants and system integrators where they are not needed. The systems are set up for one set of process operating conditions. Whenever those process conditions change, the control system vendor must come in to reconfigure the multivariable matrix. Many times the process can be controlled just as well, if not better, with conventional control strategies. The owners incur significant costs to maintain the systems, and the owner is reliant on the system vendor.

Conventional control algorithms, such as PID and dead time control, work well so long as people understand the limitations. PID controls allow a 40% change in control parameters. When tuned for the worst case, they will work over a wider range of operation. Dead time control allows only about a 10% change in parameters and will create an unstable control situation when operating outside that range.

What processes do you think are least understood and therefore not controlled correctly (pH, compressible fluids flow)?

Shinskey: pH control, especially in industry, but also in WWTP, is not well understood. The variability of the titration curve poses a significant challenge to developing effective control strategies. ORP control is also not well understood or implemented.

Hydraulic challenges, like the Deer Island Tunnel system, pose different types of control challenges resulting from the loss of the original design intent. The Deer Island Pump Station controls are into their fourth generation, and the original intent of the control philosophy was lost between the second and third generation control system. The most complex of the tunnels involves two shafts, 2 ½ miles apart, feeding in series into a deep tunnel. The wastewater flows through an 8-mile tunnel and is lifted to plant elevation by pumps. The pumps are controlled based on flow and level. Painter modeled the tunnel and determined that the shaft elevation could be controlled at either, but not both of the headworks. [The first-generation controls were developed by Painter in the early '60s and commissioned by Shinskey in the late '60s.] By the third-generation system, designed in the '80s, controls were incorrectly implemented to allow simultaneous level control of the shafts. Also, by the fourth generation the hydraulic profile was not well documented.

Another control challenge is accounting for the gain change in control loops associated with increasing and decreasing the number of controlled devices on-line. Yet another control challenge is making sure equipment is always operating in a controllable range. An example of a poor application is the recirculation valves on the gas compressors at Deer Island. The recirc valves should have opened when the compressor stepped up to the next stage and closed when the compressor stepped down. That type of logic was never incorporated into the compressor design. A good example of sequencing equipment was implemented by EMA. Looking at operating efficiency of very large pumps (3500 hp), they determined the optimum operating range for the pumps and started and stopped pumps based on demand so that all the pumps running are always running in their optimum operating range.

System retrofits pose an operational challenge: operators are accustomed to controlling the process based on prior operating instruction and experience, which may include manual operation. In order to overcome the mindset of "that's the way it's always been done," sometimes loops are reconfigured to control a different valve or pump. Changing the loop effectively erases the operating experience, requiring operators to run the facility based on the new controls.

Summary and Follow-up

In general, Shinskey commented that not much in terms of detailed control strategies is transferable from industry to WWTP. The control challenges in industrial control are process and application specific and are solved on a case-by-case basis. The places where WWTP may most benefit from industry's experience come from using methodologies developed in industry.

General points made by Shinskey related to industry experience:

Never justify an automation system based on staff reduction. Always base it on savings in energy use or improved quality control, even though part of the results from successful automation would be the operators' span of control increasing from 40 loops to 100 loops to 400 loops or more.

Most often in his experience poor design contributes to loops not working in automatic. The loop gain is wrong, etc. This would cover problems associated with oversized equipment.

When retrofitting a plant, particularly from manual to automatic, all the old controls need to be taken away in order for the operators to accept the new way of running a process. Not doing this makes it possible for operators to fall back on the old way of doing things, putting the new system at risk of not being accepted and used.

WWTP operators need to learn to operate at the permit limits. Running at discharge levels well below the permit, just because the plant can do it, costs a lot of money.

In petroleum and chemical projects the process designers and the I&C designers work together in a team from the beginning of the project. Shinskey has observed in WW that process design does not include I&C as an integral part of the job. I&C designers are low on the pecking order.

Future or follow-up, contributions may best be in developing methodologies for control implementation.

REFERENCES

Brewer, H.M.; Stephenson, J.P.; Green, D. *Plant Optimization Using On-line Phosphorus Analyzers and Automated SRT Control.* CH2M Hill Gore & Storrie Limited. Prepared for the Town of Collingwood, April 1996.

Chapman, D. T. The influence of process variables on secondary clarification. *Journal Water Pollution Control Federation* **1983,** 55, 1425-1434.

Collins, M.A.; Crosby, R. M. *Impact of Flow Variations on Secondary Clarifier Performance.* 53rd Annual WPCF Conference, Las Vegas, Nevada, Sep 28 – Oct 3, 1980.

Davison, G. Exploit advanced technologies for predictive maintenance. *Intech* **2000,** August, 62-66.

Flanagan, M. J.; Bracken, B.D. *Design Procedures for Dissolved Oxygen Control of Activated Sludge Processes.* EPA 600/2-77-032; June, 1977.

Garrett, M.T., Jr. Hydraulic control of activated sludge growth rate. *Sewage and Industrial Wastes* **1958,** 30, 253-261.

Gillette, R. A.; Joslyn, D. S. *Thickening and Dewatering Processes: How to Evaluate and Implement and Automation Package.* WERF Project 98-REM-3 – Report #D-13006, 2002.

Grace, J.; Shinskey, F. G.; Michnovez, E. *Integrating a Distributed Control System into the Existing Deer Island North Main Pumping Station.* WEFTEC '98 presentation, 1998.

Hill, R. D., *Dynamics and Control of Solids-Liquid Separation in the Activated Sludge Process.* Ph.D. Dissertation, Rice University, Houston, TX, 1985.

Hill, R. D. *Technical and Psychological Obstacles for Implementation of Automation in the Wastewater Industry.* WEFTEC '99 workshop presentation, 1999.

Instrument Society of America. *Standards and Recommended Practices for Instrumentation and Control, S5.1 Instrumentation Symbols and Identification* (updated annually).

Instrument Society of America. *Standards and Recommended Practices for Instrumentation and Control, S5.3 Graphic Symbols for Distributed Control/Shared Display Instrumentation, Logic and Computer Systems* (updated annually).

Instrument Society of America. *Standards and Recommended Practices for Instrumentation and Control, S5.5 Graphic Symbols for Process Displays* (updated annually).

Instrumentation Testing Association. *Online Ammonia Analyzers for Water and Wastewater Treatment Applications.* 2002.

Instrumentation Testing Association. Investigating sampling systems for online instruments. *ITA Analyzer* **2001,** 4, 2, 3-11.

Instrumentation Testing Association. *On-Line Monitoring to Control Transients in Wastewater Treatment.* WERF Project 92-OPW-1, 1994.

Instrumentation Testing Association. *Maintenance Benchmarking Study–Total and Free Chlorine Residual Analyzers Online.* 1999.

Instrumentation Testing Association. *Maintenance Benchmarking Study–pH Analyzers Online.* 1999.

Instrumentation Testing Association. *Maintenance Benchmarking Study–Suspended Solids and Turbidity Analyzers Online.* 1999.

Instrumentation Testing Association. *Maintenance Benchmarking Study–Instrumentation, Control and Automation Staffing.* 1999.

Instrumentation Testing Association. *Evaluation Report PER88DO-003 Model 6320 DO Monitor and Model 20 Sensor Enterra Instrumentation Technologies, Inc.* Feb, 1988.

International Association on Water Pollution Research. *Instrumentation and Automation,* Workshop Proceedings, London, England, 1973.

Kent, B. *Polymer Conditioning Improvements using Modern Technology Reduces Costs and Increases Efficiency at the Duffin Creek Water Pollution Control Plant.* Proceedings of the 2000 WEAO Technical Conference and Exposition, Hamilton, Ontario, Apr 2001.

Laquidara, M.; Garrity, G.; Tyler, C.; Waitt, W.; White, N. Water testing the secondary treatment facilities at MWRA's Deer Island Wastewater Treatment Plant. *Journal of the New England Water Environment Association* **1998,** 32, 2.

Newbigging, M. L. *Control of Pump Station Effluent Flow Rate.* Master's Thesis, McMaster University, Hamilton, Ontario, 1987.

McMillan, G.K.; Mertz, G.E.; Trevathan, V.L. Troublefree instrumentation. *Chem. Eng.* November 1998.

Olsson, G. *Automatic Control in Combined Wastewater Treatment Plants.* International Environmental Colloquim, Liege, Belgium, 1978.

Pflanz, P. Performance of (activated sludge) secondary sedimentation basins. *Advances in Water Pollution Research* **1969**, 569-581.

Roesler, J. F. *Current Status of Research in Automation of Wastewater Treatment Plants in the United States.* U.S. EPA Municipal Environmental Research Laboratory, Cincinnati, Ohio, 1977.

Rosemount Analytical. *Diagnostics and Reliability Based Maintenance.* http://www.rosemount.com/products/ams/rb-maint.htm, May 8, 1998.

Sanders, F. F. Watch out for instrument errors. *Chem. Eng. Prog.*, July 1995.

Shinskey, F. G. Control of a resonant process: The Deer Island wastewater tunnels. *Control*, October 1995, 45-51.

Stire, T. G. *Process Control Computer Systems Guide for Managers*. Butterworth Publishers: Boston, 1983, p 170.

Vitasovic, Z. *Automation in Wastewater Conveyance and Treatment Systems*. WEFTEC 2001 workshop presentation, 2001.

Water Environment Federation. *Instrumentation in Wastewater Treatment Facilities*. 1993.

Water Environment Federation. *Computers in the Water Environment*. Specialty Conference Proceedings, 1993.

Water Environment Federation. *Automating To Improve Water Quality*. Specialty Conference Proceedings, 1995.

Water Environment Federation. *Computer Technologies for the Competitive Utility*. Specialty Conference Proceedings, 1997.

Water Environment Federation. *Automated Process Control Strategies*. 1997.

Water Pollution Control Federation. *Instrumentation in Wastewater Treatment Plants, Manual of Practice No. 21*, 1978.

Water Pollution Control Federation. *Process Instrumentation & Control Systems, Manual of Practice No. OM-6*, 1984.

GLOSSARY

Analog/Digital Controller	A single loop controller; a digital implementation of an analog controller
DCS	Distributed Control System – control device consisting of processor, memory, and I/O capabilities
ERP	Enterprise Resource Planning - current evolution of manufacturing resources planning (MRP & MRPII) systems. ERP is being positioned as the foundation and integration of enterprise-wide information systems. Such systems will link together all of a company's operations including human resources, financials, manufacturing, and distribution as well as connect the organization to its customers and suppliers.
Ethernet	A local-area network (LAN) architecture developed by Xerox Corporation in cooperation with DEC and Intel in 1976. Ethernet uses a bus or star topology and supports data transfer rates of 10 Mbps. The Ethernet specification served as the basis for the IEEE 802.3 standard, which specifies the physical and lower software layers. Ethernet uses the CSMA/CD access method to handle simultaneous demands. It is one of the most widely implemented LAN standards.
FDDI	Fiber Distributed Data Interface, a set of ANSI protocols for sending digital data over fiber optic cable. FDDI networks are token-passing networks, and support data rates of up to 100 Mbps (100 million bits) per second. FDDI networks are typically used as backbones for wide-area networks.
Fieldbus	Fieldbus is a generic name given to fully-digital communication protocols for industrial measurement and control applications. Ideally, the cables used are more closely related to communication technology than to process instrumentation, but operation is also possible over conventional process field cabling, at least over limited distances (say up to 400 m.).
GC	Gas chromatograph is an instrument that partitions a sample into its separate components by selectively sorbing and desorbing onto a media and then sensing the individual components as they pass a detector.
HART	Highway Addressable Remote Transducer – digital communication protocol designed for industrial process measurement applications. Signal uses a low-level modulation superimposed on the standard 4-to-20 mA current loop. Because the HART signal is small, and composed of sine waves, its average value is zero and does not significantly affect the accuracy of the analog current signal, which can therefore still be used. This provides compatibility with existing systems, while allowing simultaneous digital communication for device configuration, status checking, diagnostics and so forth.

HMI	Human-machine interface is display software that displays information from a control system in a convenient format for people to use.
IAWPR	International Association on Water Pollution Research, now the International Water Association (IWA)
ISA	Instrument Society of America – A professional organization that sets many standards for the use of instrumentation and provides instrumentation training and certification
ITA	Instrumentation Testing Association – A not-for-profit organization of wastewater plant owners, consultants, regulators, and suppliers who pool their resources to improve the use of instrumentation in water and wastewater treatment
IWA	International Water Association – An international professional organization for the water and wastewater industries
I/O	INPUT/OUTPUT - The process of transferring data to and from a computer-controlled system using its communication channels, operator interface devices, data acquisition devices, or control interfaces.
LEL	Lower Explosion Limit – An instrument measurement for potential explosive mixtures of gasses (0-100%)
MIS	Management Information System or management information services, and pronounced as separate letters, MIS refers to a class of software that provides managers with tools for organizing and evaluating their department. Within companies and large organizations, the department responsible for computer systems is sometimes called the MIS department. Other names for MIS include IS (Information Services) and IT (Information Technology).
MOP	Manual of Practice – A manual published by the Water Environment Federation as a practical guide to a field in environmental engineering.
PID	Proportional-Integral-Derivative defines the most commonly implemented controller used to control analog control loops
PLC	Programmable Logic Controller – control device consisting of processor, memory, and I/O capabilities used to interface with and control a process
SRT	Solids Retention Time or Sludge Age – The average time a solid particle remains in the activated sludge system. SRT has proven to be an important control parameter.

TOD	Total Oxygen Demand – An instrumental measurement of the total amount of oxygen required to oxidize the organic material in a sample
Y2K	The Year 2000 – Many computer systems had problems distinguishing between the years 2000 and 1900 due to programming associated with dates
UV	Ultraviolet Radiation – usually used in wastewater disinfection
WEF	Water Environment Federation – A professional organization of the wastewater industry
WERF	Water Environment Research Foundation – water quality research organization affiliated with WEF.

2002 WERF Subscribers

The following agencies and corporations are WERF's charter stormwater research supporters:

Stormwater Agency

California
Monterey, City of
Santa Rosa, City of

Colorado
Boulder, City of

Georgia
Griffin, City of

Kansas
Overland Park, City of

Kentucky
Louisville & Jefferson County Metropolitan Sewer District

North Carolina
Cary, Town of
Charlotte, City of/Mecklenburg County

Pennsylvania
Philadelphia, City of

Corporate

BaySaver Inc.
Vortechnics Inc.
WWETCO, LLC

State

Fresno Metropolitan Flood Control District, Calif.
Urban Drainage & Flood Control District, Colo.

Wastewater Utility

Alabama
Montgomery Water Works & Sanitary Sewer Board

Alaska
Anchorage Water & Wastewater Utility

Arizona
Gila Resources
Glendale, City of, Utilities Department
Mesa, City of
Phoenix Water Services Department
Pima County Wastewater Management

Arkansas
Little Rock Wastewater Utility

California
Central Contra Costa Sanitary District
Contra Costa Water District
Crestline Sanitation District
Delta Diablo Sanitation District
Dublin San Ramon Services District
East Bay Municipal Utility District
El Dorado Irrigation District
Escondido, City of
Fairfield-Suisun Sewer District
Irvine Ranch Water District
Las Virgenes Municipal Water District
Lodi, City of
Los Angeles, City of
Los Angeles County, Sanitation Districts of
Orange County Sanitation District
Palo Alto, City of
Riverside, City of
Sacramento Regional County Sanitation District
San Diego Metropolitan Wastewater Department, City of
San Francisco, City & County of
San Jose, City of
Santa Barbara, City of
Santa Rosa, City of
South Bayside System Authority
South Orange County Wastewater Authority
Union Sanitary District

Colorado
Boulder, City of
Colorado Springs, City of
Littleton/Englewood Water Pollution Control Plant
Metro Wastewater Reclamation District, Denver

Connecticut
New Haven, City of, WPCA

District of Columbia
District of Columbia Water & Sewer Authority

Florida
Broward, County of
Fort Lauderdale, City of
Gainesville Regional Utilities
JEA
Kissimmee, City of, Department of Water Resources
Miami-Dade Water & Sewer Authority
Orange County Utilities Department
Orlando, City of
Reedy Creek Improvement District
Seminole County Environmental Services
St. Petersburg, City of
Stuart Public Utilities
Tallahassee, City of
Tampa, City of
West Palm Beach, City of

Georgia
Atlanta Wastewater Services
Augusta, City of
Clayton County Water Authority
Cobb County Water System
Columbus Water Works
Fulton County
Gwinnett County Department of Public Utilities
Macon Water Authority
Savannah, City of

Hawaii
Honolulu, City and County of

Illinois
American Bottoms Wastewater Treatment Plant
Dupage, County of
Greater Peoria Sanitary District
Metropolitan Water Reclamation District of Greater Chicago
Wheaton Sanitary District

Indiana
Fort Wayne, City of
Indianapolis, City of

Iowa
Ames, City of
Des Moines Metro Wastewater Reclamation Authority

Kansas
Johnson County Unified Wastewater Districts
Unified Government of Wyandotte County/Kansas City, City of

Kentucky
Louisville & Jefferson County Metropolitan Sewer District

Louisiana
Sewerage & Water Board of New Orleans

Maine
Bangor, City of

Maryland
Anne Arundel County Bureau of Utility Operations
Howard County Department of Public Works
Washington Suburban Sanitary Commission

Massachusetts
Boston Water & Sewer Commission
Massachusetts Water Resources Authority
Upper Blackstone Water Pollution Abatement District

Michigan
Ann Arbor, City of
Detroit, City of

Grand Rapids, City of
Holland Board of Public Works
Lansing, City of
Owosso Mid-Shiawassee County WWTP
Saginaw, City of
Wayne County Department of Environment
Wyoming, City of

Missouri
Independence, City of
Kansas City Missouri Water Services Department
Little Blue Valley Sewer District
Metropolitan St. Louis Sewer District

Nebraska
Lincoln Wastewater System

Nevada
Henderson, City of

New Jersey
Passaic Valley Sewerage Commissioners

New York
New York City Department of Environmental Protection

North Carolina
Cary, Town of
Charlotte/Mecklenburg Utility Department
Durham, City of
Metropolitan Sewerage District of Buncombe County
Orange Water & Sewer Authority

Ohio
Akron, City of
Butler County Department of Environmental Services
Columbus, City of
Metropolitan Sewer District of Greater Cincinnati
Northeast Ohio Regional Sewer District

Oklahoma
Tulsa, City of

Oregon
Clean Water Services
Eugene/Springfield Water Pollution Control
Water Environment Services

Pennsylvania
Philadelphia, City of
University Area Joint Authority, State College

South Carolina
Charleston Commissioners of Public Works
Mount Pleasant Waterworks & Sewer Commission
Spartanburg Sanitary Sewer District

Tennessee
Cleveland, City of
Knoxville Utilities Board
Murfreesboro Water & Sewer Department
Nashville Metro Water Services

Texas
Amarillo, City of
Austin, City of
Dallas Water Utilities
Denton, City of
El Paso Water Utilities
Fort Worth, City of
Gulf Coast Waste Disposal Authority
Houston, City of
San Antonio Water System
Trinity River Authority

Utah
Salt Lake City Corporation

Virginia
Alexandria Sanitation Authority
Arlington, County of
Fairfax County Virginia
Hampton Roads Sanitation District
Henrico, County of
Hopewell Regional Wastewater Treatment Facility
Loudoun County Sanitation Authority
Prince William County Service Authority
Richmond, City of

Washington
Edmonds, City of
Everett, City of
King County Department of Natural Resources
Seattle Public Utilities

Wisconsin
Green Bay Metro Sewerage District
Madison Metropolitan Sewerage District
Milwaukee Metropolitan Sewerage District
Racine, City of
Sheboygan Regional Wastewater Treatment
Wausau Water Works

Australia
Sydney Water Corporation

Canada
Toronto, City of, Ontario

Mexico
Servicios de Agua y Drenaje de Monterrey, I.P.D.

United Kingdom
Yorkshire Water Services Limited

Corporate

The ADVENT Group Inc.
Alan Plummer & Associates
Alpine Technology Inc.
Aquateam–Norwegian Water Technology Centre A/S
Black & Veatch
Boyle Engineering Corporation
BPR CSO
Brown & Caldwell
Burns & McDonnell
The Cadmus Group
Camp Dresser & McKee Inc.
Carollo Engineers Inc.
Carpenter Environmental Associates Inc.
CDS Technologies Inc.
Chemtrac Systems Inc.
CH2M HILL
Clancy Environmental Consultants Inc.
Damon S. Williams Associates, LLC
Earth Tech Inc.
Ecolab Water Care Services
EMA Inc.
The Eshelman Company Inc.
Finkbeiner, Pettis, & Strout Inc.
Frontier Geosciences Inc.
ftn Associates Inc.
Gannett Fleming Inc.
GE Betz
Golder Associates Inc.
Greeley and Hansen LLC
The HACH Company
Hazen & Sawyer, P.C.
HDR Engineering Inc.
HNTB Corporation
HydroQual Inc.
Insituform Technologies Inc.
Institute for Environmental Technology & Industry, Korea
Jacobson Helgoth Consultants Inc.
Jason Consultants Inc.
Jordan, Jones, & Goulding Inc.
KCI Technologies Inc.
Kelly & Weaver, P.C.
Kennedy/Jenks Consultants
Komline Sanderson Engineering Corporation
Lawler, Matusky & Skelly Engineers, LLP
Limno-Tech Inc.
Lombardo Associates Inc.
Malcolm Pirnie Inc.
McKim & Creed
MWH
Odor & Corrosion Technology Consultants Inc. (OCTC)
ONDEO Degremont Inc.
PA Government Services Inc.
Parametrix Inc.
Parsons
Post, Buckley, Schuh & Jernigan
The RETEC Group
R.M. Towill Corporation
Royce Technologies
Sear-Brown
SYNAGRO
Tetra Tech Inc.
Trojan Technologies Inc.
URS Corporation
USFilter NATC
Wade-Trim Inc.
Weston Solutions Inc.
Woodard & Curran
Woodruff & Howe Environmental Engineering Inc. (WHEE)
WRc/D&B, LLC

Industry

American Electric Power
BP-Amoco Corporation
Chevron Research & Technology Company
The Coca-Cola Company
Dow Chemical Company
DuPont Company
Eastman Chemical Company
Eastman Kodak Company
Merck & Company Inc.
ONDEO Services
Procter & Gamble Company
Reliant Energy
Severn Trent Services Inc.
Shell Global Solutions
Thames Water Plc
United Water Services LLC

Board of Directors

Research Council

Stormwater Technical Advisory Committee

WERF Product Order Form

As a benefit of joining the Water Environment Research Foundation, subscribers are entitled to receive one complimentary copy of all final reports and other products. Additional copies are available at cost (usually $10). To order your complimentary copy of a report, please write "free" in the unit price column. Non-subscribers may be able to order WERF publications through WEF (www.wef.org) or IWAP (www.iwap.co.uk). Visit WERF's website at www.werf.org for details.

WERF keeps track of orders. If the charge differs from what is shown here, we will call to confirm the total before processing.

Name _______________ Title _______________

Organization _______________

Address _______________

City _______________ State _______ Zip Code _______ Country _______

Phone _______________ Fax _______________ Email _______________

Stock #	Product	Quantity	Unit Price	Total
			Postage & Handling	
			VA Residents Add 4.5% Sales Tax	
			Canadian Residents Add 7% GST	
			TOTAL	

Method of Payment: (All orders must be prepaid.)

❑ Check or Money Order Enclosed

❑ Visa ❑ Mastercard ❑ American Express

Account No. _______________ Exp. Date _______

Signature _______________

Shipping & Handling:

Amount of Order	United States	Canada & Mexico	All Others
Up to but not more than:	Add:	Add:	Add:
$20.00	$5.00*	$8.00	50% of amount
30.00	5.50	8.00	40% of amount
40.00	6.00	8.00	
50.00	6.50	14.00	
60.00	7.00	14.00	
80.00	8.00	14.00	
100.00	10.00	21.00	
150.00	12.50	28.00	
200.00	15.00	35.00	
More than $200.00	Add 20% of order	Add 20% of order	

*minimum amount for all orders

Note: Please make checks payable to the Water Environment Research Foundation.

To Order (Subscribers Only):

Online

Log on to www.werf.org and click on the "Product Catalog."

Phone/Fax

Contact WERF at (703) 684-2470 or fax this form to (703) 299-0742.

Mail

Complete and mail this form to:
WERF
Attn: Subscriber Services
601 Wythe Street
Alexandria, VA 22314-1994